2012
Extinction or Utopia

About the Author

A native of Minnesota but a resident of Colorado since 1969, Jeffrey Allan Danelek's life has been a journey that has taken him down many different paths. Besides writing, his hobbies include art, political and military history, religion and spirituality, numismatics (coin collecting), paleontology, astronomy (and science in general), and Fortean subjects such as Bigfoot, UFOs, and things that go bump in the night. His personal philosophy is that life is about learning and growing, both intellectually and spiritually, and that is the perspective with which he approaches each project he undertakes. Jeff currently resides in Lakewood, Colorado, with his wife, Carol, and their two sons.

2012

Extinction or Utopia

Doomsday Prophecies Explored

J. Allan Danelek

Llewellyn Publications
Woodbury, Minnesota

2012 Extinction or Utopia: Doomsday Prophecies Explored © 2009 by J. Allan Danelek. All rights reserved. No part of this book may be used or reproduced in any manner whatsoever, including Internet usage, without written permission from Llewellyn Publications, except in the case of brief quotations embodied in critical articles and reviews.

First Edition
First Printing, 2009

Cover art © Photodisc Spacescapes
Cover design by Lisa Novak
Editing by Brett Fechheimer
Llewellyn is a registered trademark of Llewellyn Worldwide, Ltd.

Library of Congress Cataloging-in-Publication Data
Danelek, J. Allan, 1958–
  2012 : extinction or utopia : doomsday prophecies explored / J. Allan Danelek. —1st ed.
     p. cm.
  ISBN 978-0-7387-1464-6
  1. Two thousand twelve, A.D. 2. Twenty-first century—Forecasts. 3. Prophecies. 4. Judgment Day. 5. Catastrophical, The. 6. Social prediction. 7. Extinction (Biology) 8. Utopias. I. Title. II. Title: Two thousand twelve. III. Title: Twenty twelve.
  CB161.D28 2009
  133.3—dc22
                          2009021166

Llewellyn Worldwide does not participate in, endorse, or have any authority or responsibility concerning private business transactions between our authors and the public.

All mail addressed to the author is forwarded but the publisher cannot, unless specifically instructed by the author, give out an address or phone number.

Any Internet references contained in this work are current at publication time, but the publisher cannot guarantee that a specific location will continue to be maintained. Please refer to the publisher's website for links to authors' websites and other sources.

Llewellyn Publications
A Division of Llewellyn Worldwide, Ltd.
2143 Woodvale Drive, Dept. 978-0-7387-1464-6
Woodbury, Minnesota 55125-2989, U.S.A.
www.llewellyn.com

Printed in the United States of America

Other Books by J. Allan Danelek

The Case for Ghosts: An Objective Look at the Paranormal

Mystery of Reincarnation: The Evidence & Analysis of Rebirth

Atlantis: Lessons from the Lost Continent

UFOs: The Great Debate: An Objective Look at Extraterrestrials, Government Cover-Ups, and the Prospect of First Contact

Contents

Introduction 1

Chapter One
Prophecy in Retrospect 5

Chapter Two
Why We Believe 19

Chapter Three
Failed Doomsday Prophecies of the Past 29

Chapter Four
Nostradamus: The Seer of St. Rémy 45

Chapter Five
Edgar Cayce: The Sleeping Prophet 57

Chapter Six
2012 Hysteria 69

Chapter Seven
Prophecy in the Bible 83

Chapter Eight
How End-Times Prophecy Works 91

Chapter Nine
More Dispensationalist Meanderings from the Bible 101

Chapter Ten
The Antichrist 121

Chapter Eleven
Real Doomsday Scenarios 133

Chapter Twelve
Ecological Doomsday 157

Chapter Thirteen
The Extraterrestrial Hypothesis 167

Chapter Fourteen
The Problem with Foretelling the Future 175

Chapter Fifteen
Extinction or Utopia: What the Future May *Really* Hold 183

Conclusion 195

Introduction

According to the Mayan calendar, the world as we know it will end sometime around December 21, 2012.

Or maybe it won't.

Of course, if the world doesn't end then, that won't mean we'll be safe from destruction. According to biblical prophecy, Jesus Christ could return at any moment to destroy the armies of the Antichrist and reestablish his throne in Jerusalem, thereby ushering in a thousand years of peace. However, if that doesn't occur, there's always the chance that the Mahdi will arise to institute a kingdom of justice and, alongside the returned Isa Al-Maseeh (Jesus), fight against the Dajjal, the Antichrist of Islam. Then, of course, there's always the chance that—at least according to the Hopi Indians—a blue star will suddenly appear in the sky to signal the start of a great atomic war that will destroy the white man and other ancient races.

However, it's not just religious end-times prophecies we need to worry about. Doomsday prophecies also come in secular and even scientific packages as well, as evidenced by an ever-growing influx of climatologists, futurists, scientists, and environmentalists all jumping onto the doomsday bandwagon to warn us about their own catastrophic scenarios: global warming will wipe out entire species and obliterate ancient ecosystems; unlimited population growth will lead to social unrest and political chaos, resulting in the collapse of whole civilizations; the next great pandemic will leave millions—if not billions—dead, possibly spelling the extinction of the human race; a thermonuclear war will turn the planet into a glowing, uninhabitable radioactive cinder; an asteroid, comet, giant meteor, burst of cosmic radiation, massive volcanic eruption—you name it—will snuff out life on the planet. It seems that the list of things that are going to destroy us is growing longer and more gruesome each day.

If these were merely the beliefs of small fringe groups or environmental extremists, they probably wouldn't constitute that much of a problem. However, doomsday beliefs are becoming mainstream—their influence being seen not only on bookstore shelves and in the pseudoscientific docudramas that seem to permeate the cable channels, but even finding their way into government legislation designed to prevent us from being fried, pulverized, irradiated, or otherwise obliterated. Even credentialed scientists are getting in on the act, recounting the various doomsday scenarios available to us in frightening detail, complete with computer models and state-of-the-art computer animation.

Whether one calls it the Second Coming, Judgment Day, or the Battle of Armageddon, or whether one believes the end times will come in the guise of impending environmental disaster, global political collapse, or through some sort of nuclear doomsday, end-times prophecy remains a big part of our social identity and one that is only likely to become even bigger as we move further into the new millennium. Furthermore, in light of the apparent "funk" the planet seems to be going through nowadays, such beliefs are growing in popularity and show no signs of abating anytime soon, leading me to assume that they will probably remain a part of our fear-based culture for decades to come.

So where do these ideas come from and, more importantly, why do we continue to embrace them even today in our supposedly less superstitious and more scientific age? Why does it appear that, instead of such ideas diminishing in the harsh glare of reason, doomsday predictions are actually increasing and becoming an even more prevalent aspect of our modern culture? What drives them and gives them the fuel required to infuse our airwaves, bookshelves, and imaginations? More importantly, what does this preoccupation with our own demise say about us as a people?

It's hard to pinpoint the reason we seem so preoccupied with doomsday, but after five decades of life I think I am beginning to understand why countless millions embrace such ideas, and it is these observations I hope to explore with you in this work. I do this in the hope that we might gain a better understanding of human nature and, even more importantly, so we might be able to protect ourselves better from the damage such beliefs can produce. Doomsday predictions can leave one living beneath a cloud of pessimism that can color everything a gloomy shade of gray. That alone should be more than enough to entice us to look into the issue in some detail if only so we might disarm some of the more odious ideas before they can become self-fulfilling prophecies.

This book is also written for those who—inundated by doomsday scenarios from religious leaders, paranormal writers, New Age gurus, and politically motivated environmentalists—would just like to bring some balance to the debate. To accomplish that, we will examine the many failed prophecies humanity has believed in the past and consider how they not only influenced their cultures but also how they continue to affect their perspective on the future today in both subtle and obvious ways. And, finally, we will hold those self-proclaimed prophets of the apocalypse—both living and dead—accountable for their predictions, for it is only by taking a critical look at these people and understanding why we embrace their ideas so readily that we can come to appreciate the power they hold over us.

I hope the reader will come to recognize through all this that I am, at heart, an optimist. I don't believe for a moment that the future is as bleak and hopeless as many assume, nor do I consider hope and faith to be foolish ideals. I believe in tomorrow—however naïve that may sound—and I hope that this work in some small way will help others find the confidence and assurance that humanity, for all its many flaws, is not on the verge of its own destruction but may, in fact, be standing at the threshold of a bright and remarkable future. If this book can play even a small part in demonstrating that possibility, I will consider it to have been a worthwhile endeavor.

J. Allan Danelek
January 2009

Chapter One

Prophecy in Retrospect

Mention the year 2012 to most people, and chances are they will think of doomsday, which is remarkable considering that just ten years ago few people would have assigned any special significance to the year at all. Today, however, all one has to do is type *2012 doomsday* into Google, and the search engine will produce over 700,000 hits. Clearly, 2012 has captured the public's imagination in a way few things can.

But 2012 is nothing new. Large numbers of people have believed that other years in recent history have also portended ominous events. The most recent of these occurred during the closing months of the twentieth century, when millions looked upon the year 2000 with trepidation because of the so-called Y2K problem. Fueled by overblown media speculation, people feared that as a result of a practice in early computer program design that would cause some date-related processing to operate incorrectly on January 1, 2000, all sorts of disasters—from airliners falling from the skies to nuclear power stations melting down—would result.

This concern became so entrenched that toward the end of 1999, many people began stockpiling provisions, fearing a disruption in global food supplies, and buying generators in anticipation of worldwide electrical blackouts. Even major corporations spent billions of dollars for software programs and IT support, all designed to offset the imminent disaster.

As each country and time zone celebrated the start of the new year with scarcely a whisper of impending doom, however, it soon became clear that the belief that computers worldwide would crash at the stroke of midnight was erroneous. In the months following the great non-event, the public came to realize just how overblown and hyped the whole thing had been, and people were able to laugh off the whole affair and go on with their lives, bemused by the fact that so many others—even entire corporations—could be frightened by something as innocuous as turning the page on a calendar.

The incident did, however, serve once more to demonstrate humanity's propensity to believe even the most extraordinary claims, especially if they have the veneer of "science" to support them (many of the arguments in support of the Y2K hypothesis were compelling and apparently based on solid computer science). Clearly, the planet had been taught another lesson about how little we can really know about what the future holds for us, which would seem to be a lesson we would retain. However, as the current craze over 2012 suggests, it's apparent that we are once again preparing for the worst, demonstrating once more how very short our memory is and illustrating our apparently built-in proclivity toward embracing worst-case scenarios.

The question, however, is not only why people are so enthralled with the idea that the change of millennium might spell disaster or why 2012 has special significance for the planet, but why so many people believe that any particular date can be foretold with any degree of reliability by "prophets" or visionaries. What is it that causes so many to accept so readily such prognostications as fact, especially when often extraordinary claims are usually based on only the flimsiest of arguments?

Before we get any further into the question, it will be necessary to lay some groundwork. It has been my observation that when the subject of foretelling the future comes up—what we call prophecy—folks frequently have very different ideas of what the term means. As such, before we can examine the issue of doomsday in any detail, it will first be necessary to define what, precisely, prophecy is and how it works—for without a good, basic understanding of how the prophecy "game" is played, it will be difficult to appreciate the role such beliefs have played throughout history and the effect they continue to have today.

Defining Prophecy

The word *prophecy* comes from the Greek word *prophetes*, which means, literally, "fore-speaker" (i.e., one who speaks of an event before it occurs). In most traditions, a prophet is an intermediary between God—or the gods—and humanity, making a prophet, then, a most important figure within any culture, tribe, or religion. In fact, the prophet, being a vital link to the divine without whom the tribe would be unable to ascertain the gods' will and so potentially be led astray, was the second most important position a man or woman could hold after that of leader or tribal chief. Since being gifted with the knowledge of what God thinks, desires, or requires people to do gives one considerable power to effectively control people and, to a large degree, even dictate future political and socio-economic changes, in many cases a prophet was even more important than a king, making it not unusual for a ruler to hold authority over a people only with a prophet's blessing.

Not surprisingly, then—and conspicuously unlike today—being designated a prophet was not to be taken lightly. In the Old Testament, for example, the test of a true prophet was that *everything* he or she prophesied had to come to pass *exactly* as described or the person would be exposed as a false prophet and promptly put to death. This doubtlessly cut down considerably on the number of people claiming to be prophets and made "real" prophets exceedingly rare.[1]

However, it's important to realize that a prophet did more than sit around predicting future events and then wait to be lauded for being

---

1. There are fewer than a dozen prophets named in the Old Testament, which is remarkable considering the nearly one thousand years of history the OT texts cover.

right each time. Predicting the future was only one type of prophecy and constituted a tiny percentage of all the prophecies uttered. In other words, prophecy is not synonymous with soothsaying—that is, the ability to foretell future events—a point that is often overlooked by end-times proponents. In reality, there are three different categories or "types" of prophecy that have been identified over the centuries, each of which we will look at here.

The Introspective Prophecy

The first type of prophecy—and probably the most common—is the introspective prophecy, which is designed to provide the people with a divine perspective on the current state of affairs. An example of such a prophecy can be found in the Old Testament book of Hosea, written over 2,500 years ago:

> Yet I am the LORD thy God from the land of Egypt, and thou shall know no god but me: for there is no savior beside me. I did know thee in the wilderness, in the land of great drought. According to their pasture, so were they filled; they were filled, and their heart was exalted; therefore have they forgotten me. (Hosea 13:4–6)[2]

In other words, introspective prophecies are simply God's opinion, usually designed to remind the people what He has done for them. Though they sometimes were made in an attempt to encourage the people or strengthen their resolve, usually they were indictments against the people that point out—usually at great length and in gory detail—their lack of faith or failure to remember their place in the scheme of things. Such edicts frequently outline divine displeasure and, as such, hint at potential future consequences, but for the most part they were simply an overview of how the people found themselves in the mess they were in.

---

2. All the biblical quotes in this book are taken from the King James Version translation of the Bible, first published in 1611.

The Promissory Prophecy

Promissory prophecies, in contrast, deal less with how things are and more on how they will be in the future—either good or bad—depending upon the people's actions. An example of such a prophecy can be found in the Pentateuch (the first five books of the Old Testament and the basis for the Jewish Torah) in the twenty-eighth chapter of the book of Deuteronomy:

> *And it shall come to pass, if thou shall hearken diligently unto the voice of the LORD thy God, to observe and to do all his commandments which I command thee this day, that the LORD thy God will set thee on high above all nations of the earth. And all these blessings shall come on thee, and overtake thee, if thou shall hearken unto the voice of the LORD thy God . . . The LORD shall cause thine enemies that rise up against thee to be smitten before thy face: they shall come out against thee one way, and flee before thee seven ways. The LORD shall command the blessing upon thee in thy storehouses, and in all that thou settest thine hand unto; and he shall bless thee in the land which the LORD thy God giveth thee . . . And the LORD shall make thee plenteous in goods, in the fruit of thy body, and in the fruit of thy cattle, and in the fruit of thy ground, in the land which the LORD swore unto thy fathers to give thee.*

In essence, promissory prophecies outline specific rewards or punishments for particular activities or actions a people might take. However, while such promises can sound like predictions, they are not soothsaying in the truest sense of the word. They're actually divine intentions that may never come to pass at all if certain actions or conditions aren't met. In other words, the accuracy of such predictions is dependent upon the actions of the hearer, making such prophecies entirely conditional.

What also differentiates the promissory prophecy from the predictive prophecy is timing; promissory prophecies are of a much more immediate nature, with the reward or punishment experienced in the immediate future rather than in the years or decades to come. The reason for this is clear: prophecies that won't be realized for decades or centuries had little practical application to people for whom survival was a daily struggle, making the gods and their predictions useful only in their day-to-day lives. In other words, just as the prediction that our sun will go nova in four billion years has little practical application today, so too were predictions about events to be realized hundreds or thousands of years in the future of little concern to the ancient listener.

Predictive Prophecies

The third type of prophecy is the predictive prophecy, which is when a prophet specifically speaks of future events that are not dependent upon the actions of the people in order for them to be realized. In essence, predictive prophecies are spoken of as if they are preordained, and this is the type of prophecy most people think of when they hear the word *prophecy*. While relatively rare, examples can be found throughout the Bible, especially in the Old Testament. One such example is in the Old Testament book of Ezekiel, chapter 38:

> *And the word of the LORD came unto me, saying, Son of man, set thy face against Gog, the land of Magog, the chief prince of Meshech and Tubal, and prophesy against him . . . I will turn thee back, and put hooks into thy jaws, and I will bring thee forth, and all thine army, horses and horsemen, all of them clothed with all sorts of armour, even a great company with bucklers and shields, all of them handling swords . . . in the latter years thou shalt come into the land that is brought back from the sword, and is gathered out of many people, against the mountains of Israel . . . and thou shalt come from thy place out of the north parts, thou, and many people with thee, all of them riding upon horses, a great company, and a mighty army: And thou shalt come up against my people of Israel, as a cloud to cover the land; it shall be in the latter days, and I will bring thee*

against my land, that the heathen may know me, when I shall be sanctified in thee, O Gog, before their eyes.

Such a prophecy is clearly anticipating a future event—in this case, the invasion of Israel by someone named Gog—making it truly predictive and an example of what most people think of when discussing prophecies.

There are a couple of points to recognize with such prophecies, however, and that is that they are usually vague in nature and lack useful specifics. To cite the example above, the prophet does not clearly articulate who Gog is (though his listeners may have understood what the term meant in the context of their era) or exactly when this force was to arrive. The prophecy simply says, "in the latter years," which could mean anything from a few years to decades to even centuries in the future. In other words, such prophetic utterances usually lack the sort of precise details that would make them hold truly practical value.

Of course, the reason for this is obvious: prophets who deal with future events have to be careful about being so precise that they make it potentially possible to circumvent their prophecies through human action. In other words, were the ancient Hebrews to be told precisely who "Gog" was and when this invasion was to occur, they would have time to raise an army and defend against such an attack, potentially altering the outcome and circumventing the prophecy.

Precise prophecies could prove to be even more problematic today. By way of an illustration, let's say that the famed sixteenth-century French mystic Nostradamus had written a quatrain that clearly stated that at the end of the year nineteen hundred and forty-one, a people symbolized by "the rising sun" would attack the "Great Eagle across the sea" and sink her "mighty fleet" while it lay at anchor, thereby starting a great war that would "consume the whole world." In being so precise, it's difficult to see how world leaders in 1941 would have been able to ignore such a quatrain (or why Japan would have gone forward with her attack on Pearl Harbor if its plans had been so thoroughly compromised). We will discuss this aspect of predicting the future in more detail later, but suffice it to say that soothsaying can only be done with the greatest care lest the entire prophecy be undone by human intervention.

Being vague also makes such prophecies less dangerous to their subject. For example, if a prophet were to proclaim that a particular individual was to one day be installed as king, it might induce his predecessor to go after that individual, thereby altering history if he or she is successful in killing the future leader. (Of course, it might be argued that in such a case, God would foresee this and take it into account, but that's another issue.)

Another point to consider is that often many predictive prophecies were written *after* the fact,[3] making them actually *postdictive* rather than *predictive*. In other words, they may be historically realized events that were intentionally recorded in such a way as to make them sound as though they were predicted by God beforehand, usually for the purpose of infusing important historical events with divine significance. This is not necessarily evidence of fraud, however, but a common human tendency to see a divine hand in past historical events (in the way that some people may be tempted to reinterpret the Great Depression of the 1930s as God's punishment for the moral excesses of the 1920s). Finally, it is important to recognize that most predictive prophecies are not apocalyptic in nature, nor do they normally concern themselves with doomsday pronouncements. Most simply deal with events that are to take place at some point in the future, not at the end of time.

Doomsday Prophecies

I know I said there were only three types of prophecies, but a fourth category could be created as long as we understand it to be a subcategory of one of the three established types. This would be that tiny fraction of

---

3. This has always been a problem with the biblical narrative, since the precise date that a particular book—especially in the Old Testament—was first penned is often uncertain and frequently much later than conservative scholarship has maintained.

prophetic utterances that appear to deal specifically and undeniably with what are called the "end times." An example of such an apocalyptic pronouncement can be found in the famous Book of Revelation:

> *And he gathered them together into a place called in the Hebrew tongue Armageddon. And the seventh angel poured out his vial into the air; and there came a great voice out of the temple of heaven, from the throne, saying, It is done. And there were voices, and thunders, and lightnings; and there was a great earthquake, such as was not since men were upon the earth, so mighty an earthquake, and so great. And the great city was divided into three parts, and the cities of the nations fell: and great Babylon came in remembrance before God, to give unto her the cup of the wine of the fierceness of his wrath. And every island fled away, and the mountains were not found. (Revelation 16:16–20)*

These types of prophecies are the ones that garner the most interest among doomsday cultists and create the most problems for their believers, and as such are the specific type of prophecy we will explore in greatest detail.

However, before we can do that, it is important to understand that various doomsday scenarios see the end times in very different ways. Plus, not all of them are generated by prophets in the traditional sense (that is, a messenger from God) but can even come from secular and even scientific sources, further muddying the prophetic waters. As such, even doomsday or end-times prophecies can be further subdivided into five basic types or genres, each powered by very different belief systems.

"Genres" of Doomsday Prophecies

The most common and ancient type are faith-based predictions that find their sources in the proclamations of a particular deity or religious entity, or are derived from supernatural elements. Examples of such doomsday pronouncements include Judgment Day and Christ's return, as well as those having to do with the emergence of the Jewish Messiah or, in Islamic tradition, the Mahdi. Such beliefs, in addition to being found in numerous cultures around the world, served as the basis for almost all end-times prophecies until the twentieth century, and though they are being rapidly supplanted by non-faith-based predictions, they continue to predominate end-times thinking today and are likely to continue to do so well into the future.

A second type of doomsday prophecy—and one growing in influence today—is based around nontraditional religious belief systems. These are usually considered Eastern or "New Age" doomsday or end-times scenarios, and they find their sources not in the proclamations of God but within the complexity of pyramid geometry, crystal balls, tarot cards, astrology, and other earth religions. They remain faith-based, however, in that they base their interpretations on supernatural power that exists outside of man, and one that he may or may not have control over.

Unlike traditional faith-based prophecies, nontraditional faith-based end-times prophecies often tend to be date-specific (as the 2012 prophecies are), and so they frequently encounter major problems when the date passes uneventfully. However, on the plus side, these types of end-times predictions are often more positive in nature. In other words, they may not anticipate the end of the age to be synonymous with the destruction of the planet or its inhabitants, but instead may interpret it as being utopian in nature—something to be anticipated rather than feared. The dawn of these new golden ages are often predicted to coincide with great upheavals in the social fabric as well, so they may not always be pain-free periods either. As a rule, however, they usually point toward a more positive future for humanity rather than to its extinction, in marked contrast to most end-of-the-world prophecies.

A third and very modern type of doomsday prediction is based less on faith—though that does remain an element of it—and more on known natural processes. In effect, they are generated by science and deal with such catastrophic events as comet strikes, asteroid collisions, volcanism, pandemics, and climatic changes—among other things—as being the source of Earth's demise.

Being based on real science makes these types of predictions particularly frightening, in that they are not merely hypothetical doomsday scenarios but are repetitions of major extinction events that have proven track records of doing tremendous damage to our planet over the course of its history. Being based on observable catastrophic events in the past, then, such predictions not only portend great lethality (and, often, finality), but it also makes them, in many cases, practically inevitable. What also makes them so frightening is that the events predicted can occur almost without warning, providing little time to potentially circumvent them (if doing so is even technologically feasible). This also makes them extremely short-term predictions in contrast to faith-based predictions, which may refer to events occurring centuries or even many millennia in the future.

Additionally, such end-times predictions are highly flexible in the sense that they are often subject to modification as circumstances change or further data becomes available. In other words, the prediction that a massive asteroid will strike the planet in eight months may change once more data is accumulated and the asteroid's trajectory can be more accurately calculated, thereby making doomsday not only provisional but largely subject to change without notice. Another problem with such predictions is that they're often based on computer models, which, usually as a result of faulty or incomplete data and other unpredictable but assumed circumstances, may not be accurate. This makes such predictions extremely difficult to make with any degree of certainty, hurting science's own credibility if too many false alarms are raised.

Unlike faith-based doomsday scenarios, however, science-based scenarios often assume that some of these events can be circumvented (such as an asteroid being deflected through the use of technology), although just as often they remain beyond our control (such as the sun going nova). Also unlike faith-based doomsday scenarios that look for life to continue on after the "big day" in some greatly modified form (the "end" ushering in the advent of a utopian world or millennial kingdom, for example), science-based scenarios are usually more pessimistic and destructive, and often predict the complete extinction of humanity and even, in some cases, all life on earth.

A fourth type of prophecy falls somewhere between these two extremes and are what I call pseudoscientific prophecies. These would be doomsday predictions that are based on highly speculative theories that include such things as the earth being consumed by a micro black hole or extraterrestrials arriving either to save or conquer us. I classify them separately due to the fact that they are neither religious in nature nor particularly scientific either—or, at least, they are based on some very shaky science indeed. However, they are proving to be an increasingly popular type of doomsday prediction, largely because of their mysterious nature and, in the case of the extraterrestrial scenario, their appeal to modern sensibilities and fears.

The fifth and final type of doomsday prophecy is what I like to call social doomsday scenarios. These are predictions that don't generally anticipate grave physical danger to the planet itself—such as an asteroid strike or rapid climatic changes—but rather catastrophic change to its inhabitants in the form of social upheaval, revolution, and warfare. Overpopulation scenarios, fear of loosening moral standards, and competition for dwindling natural resources are the engines that drive these scenarios, and what makes them so attractive is that they are based, at least in part, on historical precedence, making them not entirely implausible. Fortunately, in that human nature is frequently one of the most difficult things to predict and is usually more amenable than expected to political action and social pressures to avert the danger (e.g., introducing effective birth control in an effort to curb world population growth, creating nuclear nonproliferation treaties to reduce political tensions, and so on), social apocalypses usually go unrealized.

Conclusions

The idea of doomsday, then, far from being merely a religious contrivance, appears to have been embraced by all aspects of society, including the secular and scientific communities, making each scenario more potentially credible than at any time in history. Add to the prospect of divine judgment the possibility of nuclear, environmental, societal, and cosmic destruction, and it's not difficult to understand why so many people today hold to an increasingly pessimistic view of the future. With so many things capable of destroying us, it seems that the odds are against us—or, at least, that appears to be the perspective of many people.

But that doesn't answer the question of why we believe such prophecies to be valid. Regardless of what type of end-times scenario we adhere to, what is the engine that drives and sustains our preoccupation with "doomsday" (however we choose to define it)? To answer that question, it will be necessary to examine the psychology behind end-times beliefs.

Chapter Two

Why We Believe

On the morning of March 26, 1997, San Diego police were called to a rented mansion in the upscale community of Rancho Santa Fe to investigate reports of a possible death. When the police arrived, they made a most horrific discovery: there, lying in bunk beds, dressed in identical black shirts and sweat pants, and wearing brand-new, black-and-white Nike tennis shoes, were thirty-nine rapidly decomposing bodies.

Investigators were at first baffled. There was no sign of foul play or a struggle of any kind, and none of the bodies bore any scars or evidence of trauma. In fact, it looked as though each had died peacefully in his or her sleep, making it appear that the thirty-nine men and women—ranging in age from twenty-six to seventy-two—had each taken their own lives in an act of mass suicide. This assumption was later borne out when autopsies revealed that they had all ingested a concoction of phenobarbital and vodka, and had also placed plastic bags around their heads to cause asphyxiation. Even more remarkable, they had not all died simultaneously, as is usually the norm in such cases. Instead, the suicides had been conducted in shifts over a three-day period—one group of fifteen the first day, another group of fifteen the second, and the final nine on the third day—with members of each successive group being careful to clean up and put things in order before taking their own lives.

So who were these people and why did they take their own lives—and even more mysteriously, why did they do it in such a carefully planned and complex method?

They did it because of a comet.

Actually, that's an oversimplification. They did it because they were convinced suicide was the only way their souls could hitch a ride on a spaceship they believed was hiding in the tail of Comet Hale-Bopp, then on approach to the sun and brightly visible in the night sky.

It turns out that the dead men and women were members of a group of cultists known as Heaven's Gate—a tiny group of dedicated believers who had been convinced by a former music teacher turned New Age guru, Marshall Applewhite, that planet Earth was about to be recycled and that the only chance to survive was to leave it immediately in the spaceship that rode in the comet's tail. The members of Applewhite's cult took him seriously enough to join him in taking their own lives as part of a ritualistic suicide pact.[1]

We could dismiss such behavior as the actions of largely illiterate or clearly confused people, but that would be untrue. Applewhite's followers were not ignorant or stupid. In fact, they financed their tiny cult by offering professional website development for paying clients, which demonstrates that many of them were not only intelligent but even technologically savvy. So why did they choose to believe the incoherent ravings of a man who, by nearly all accounts, was mentally unbalanced?

We may never know. However, we do know that they are not the only ones to have followed the teachings of obviously unbalanced men and women over the years. Time and time again people have believed doomsday prophets to their own demise. Of course, not all of them—or even many of them—end up taking their own lives as the

---

1. According to literature recovered at the scene, the thirty-nine victims did not consider their actions to be suicide, however. Believing that their human bodies were only vessels meant to help them on their journey, they felt that terminating their physical existence was the only means by which they could "transition" to the next, higher plane.

Heaven's Gate cult members did. That was an extreme case, much like the 913 men, women, and children who took their lives or were killed at the behest of Jim Jones and his henchmen in the jungles of Guyana in 1978,[2] but the damage such beliefs can produce can be devastating in other ways, too. For some, faith in these gurus of doom may result in financial loss, personal humiliation, and depression once their "prophet" is proven to be delusional. For others, it may mean years of living in expectation that at any moment they could be "taken," causing many to put off college, marriage, and other plans for the future in anticipation of the event, thereby inducing a kind of paranoia that can, in some cases, be debilitating. In every case, however, such faith results from a willingness to believe the most extraordinary claims made by the most unstable people.

But why do people choose to believe such claims in the first place? While a number of books have been written exploring just that subject, we can greatly condense some of the reasons and examine them here, at least briefly, in an effort to come to understand the rationale that drives doomsday prophecies and the people who not only make them, but also believe in them.

Doomsday Personality Types

There are a number of reasons why people believe doomsday prophecies. Probably the most common and, in a way, understandable, rationale has to do with our discomfort with the uncertainties of life and the feeling

---

2. Peoples Temple was a cult that believed doomsday was at hand and had moved to a compound called Jonestown in the jungles of Guyana to await "the end." Accused of holding members against their will, the group was being investigated by American authorities when the cult's leader, Reverend Jim Jones, persuaded his followers to drink cyanide-laced Flavor Aid in an effort to avoid the coming apocalypse. Nine hundred and thirteen of his followers—men, women, and children—as well as several members of the media and U.S. congressman Leo Ryan (who was leading the investigation into the charges against the group and was murdered by temple gunmen) died in the incident, making it the worst case of mass suicide in American history.

that things are out of control on our planet. We live in uncertain times, and uncertain times breed feelings of fear and foreboding in some people. Many subsequently feel trapped in a world full of things that can kill them, enslave them, and impoverish them, leaving them with the feeling that the future is not only bereft of hope, but that events are also spiraling out of control each day with ever greater velocity.

Doomsday scenarios, then, not only confirm this perception, but in an almost contradictory way, then turn around and offer people hope that things aren't always going to remain as chaotic as they appear and that there is a plan and purpose behind all the insanity. To many—especially those who embrace messianic end-times beliefs—such scenarios provide a sense that someone, presumably God, is in control and that He will ultimately rescue His creation and initiate the paradise on Earth many so desperately crave. Doomsday prophecies, then—at least those that promise some sort of utopian aftermath—are the ultimate "happy ending" and, for many people, a source of considerable hope and comfort. (Even if they must endure a period of tribulation, if heaven is the end product it is worth the price in many people's minds.) Unfortunately, this type of belief system is also highly contagious and capable of attracting large number of followers if perpetuated by people who are particularly convincing.

Another reason some people are so attracted to doomsday beliefs is because such beliefs frequently describe exciting and fantastic events that offer relief from the monotony of life. In effect, they are escapist fantasies—a promise, if you will, that the banalities of the world we inhabit are only temporary and that they will one day be punctuated by a truly remarkable series of events. (And what could be more spectacular than Armageddon and the Second Coming of Christ?) Boredom is a powerful incentive to believe the unbelievable, if only as a distraction from the ordinariness of life.

Another group susceptible to embracing doomsday scenarios are those who see the earth as such a terrible place that its destruction might be interpreted as a positive turn of events. These are the professional cynics who so fixate on the negative circumstances of life that the promise of complete destruction simply confirms to them that the world is, indeed, a particularly nasty place, thereby justifying their fear and, the truth be told, their cynicism. In fact, many who live in such a world are often annoyed when you suggest that things aren't that bad, as if optimism were irrational and pessimism the only recourse for humanity. Fortunately, such people tend to repel rather than attract converts, thereby usually limiting the impact they have on others.

But perhaps the strongest incentive for embracing doomsday scenarios is the smugness that people get from being privy to the future—to feeling that they are "in" on some great cosmic secret. It's almost as if they and a select group of "chosen ones" have been given a special opportunity to peek behind the curtain of eternity as a reward for their cleverness in figuring out the puzzle. For people whose lives are a bit on the ordinary side, such a faith structure brings color to a drab existence and imbues it with a meaning and purpose it previously lacked.

This need to maintain the illusion of self-importance and infallibility often makes such people impervious to being dissuaded from their beliefs, no matter how many times their prophet of choice has been proven wrong or how thoroughly the weight of evidence is stacked against them. Doomsday believers frequently believe with a fanaticism bordering (and often crossing) the line between faith and irrationality, making them particularly vulnerable to coercion and manipulation, especially when their leader possesses an extraordinary degree of charisma combined with an uncanny ability to convince others of his or her claims. That combination, when merged with a naïve willingness to suspend disbelief and an unwavering allegiance to a leader and/or belief system, explains how a Heaven's Gate and a Jonestown are possible. Faith is a wonderful thing when invested in something noble, but like a two-edged sword, it can prove a deadly tool in the hands of the wrong person.

Lack of Critical Thinking Skills

But that doesn't entirely answer the question as to why intelligent men and women—many of whom often hold PhDs—are so quick to accept the most outrageous claims as a matter of course and hold to them with such tenacity. It seems that common sense and rationality would quickly expose the fallacious nature of most doomsday claims and limit their adherents to only the most fringe elements of society, yet we often see such beliefs evident within large segments of mainstream America. How can this be?

The reason has to do, I think, with the fact that many people simply lack the critical thinking skills so vital in determining fact from fiction. I know this sounds harsh, but the truth is that many people—I am tempted to say most—prefer to accept ideas on a purely intuitive level rather than on a rational one. As such, for many, end-times prophecies make perfect sense not because their proponents have in any way laid out a good, logical case for their beliefs or have demonstrated an impressive track record of making accurate predictions in the past, but because humans often "feel" something to be true. As is often seen in the political arena—especially during an election cycle—it is perception that frequently trumps reality, with people accepting things as being "good" or "bad" based not on facts but on rhetoric and hyperbole, thereby short-circuiting the ability to determine the truth with any degree of accuracy or consistency.

Additionally, most people are not normally predisposed to acquire the knowledge base necessary to recognize a fallacious statement when they hear one. For the most part, the average man or woman lacks the historical or scientific knowledge necessary to recognize the validity of a specific prediction or possess a context within which to weigh the validity of a particular hypothesis. For example, how many have ever taken the time to study failed prophecies of the past in an effort to acquire some perspective on the subject, or really understand how likely our planet is to be struck by a killer asteroid in their lifetime? And when it comes to Bible-based end-times prophecies, how many professing Christians have ever studied the various competing theories held by theologians throughout history in an effort to weigh the merits of each position? And, finally, how many people have a good working knowledge of ancient history and so can determine whether a particular prophecy may have already been realized in historical events of the past? Unfortunately, it takes work to see through the fallacies, suppositions, historical inaccuracies, and just plain nonsense that are such a major element of end-times scenarios, which is something most people have neither the time nor the inclination to do.

That most people lack a solid historical, scientific, or rational basis upon which to form their opinions is what also makes them prone to defer to the opinions of others—especially if they believe that the said others are experts of some kind or endowed with a mantle of spiritual authority. As such, if an environmentalist claims that the world will run out of energy in twenty years or a televangelist enthuses about how the headlines are demonstrating that Jesus could return at any time, many believe them without question—often without the evidence to support their claims or without considering counterarguments. In effect, we are often simply too willing to accept other people's words and interpretations as fact, which can be a dangerous thing to do.

But shouldn't we defer to those who might know more than we do? Certainly, a scientist who warns us that the earth is warming at an alarming rate doubtlessly knows more than the average man on the street, so wouldn't we be wise to take such a person seriously? And theologians who have spent a lifetime studying ancient texts are clearly in a better position to interpret these writings than a mere layman would be, so wouldn't we be wrong to ignore them?

Obviously we need to listen to professional scientists and should respect the teachings of credentialed biblical scholars, but these are not normally the people causing all the problems. It is the self-taught experts who usually make the most remarkable predictions; real scientists and theologians are normally too cautious to boldly set dates or write books recounting doomsday in detail. It is the eccentric, the obsessed, the presumptuous, and the just plain reckless who make most of the boldest and outrageous claims. Paradoxically, it sometimes appears that the very outrageousness of their claims is what makes their ideas so attractive in the first place, making their success even more inexplicable.

In the end, we must learn to hold such people accountable for their predictions, both when they make them and when the predicted events fail to materialize as promised. Clearly, had the followers of

Marshall Applewhite's Heaven's Gate cult simply stepped back and examined the man's claims from a scientific/rational perspective, they would have quickly seen the absurdity behind his statements and either challenged him on them or left the group (as a few had done in the years before the mass suicide). So why didn't they?

Because they chose to give Applewhite intellectual authority over themselves rather than accept responsibility for their own beliefs, making them especially susceptible to accepting the most absurd claims without question regardless of their level of intelligence, social status, or educational level. Theirs was a world that existed in black and white without shades of gray and one that worshiped a type of certainty that is unwavering even in the light of reason, science, or subsequent events. Such a mindset is what makes it possible for a cult to convince its followers to commit mass suicide or, for that matter, for any fantastic claim to be taken as truth by large numbers of people. Evidence is not what matters here, but faith. Trust becomes a more valuable commodity than truth, and conformity to the group mindset becomes a virtue to be pursued with unceasing devotion. In effect, belief trumps knowledge, leading many a sincere and otherwise honest person to embrace the most outrageous premise and suffer as a consequence of doing so.

Unfortunately, there are any number of people quite willing and capable of playing to those feelings. Marshall Applewhite, Jim Jones, David Koresh, Charles Manson, and scores of others like them (along with the likes of Adolf Hitler, for that matter) knew how to play upon human nature and manipulate people into doing things they would never otherwise have considered on their own. It's always impressed me what people are willing to believe if the one making the claims appears sincere, intelligent, and dogmatic, regardless of how nonsensical those claims may be. Additionally, many people lack the ability to recognize when they are in the presence of a truly disturbed or fantasy-prone personality, regardless of how obviously disturbed their leader may appear to the objective observer. This inability to discern mental illness—extraordinarily evident in the cases of Marshall Applewhite and Reverend Jim Jones—when combined with a fervent desire or need to believe, can be a fatal combination.

Of course, that's not to imply that all people who make doomsday predictions are evil or consciously aware that they are manipulating others. Some truly do believe their own rhetoric and happen to possess the sort of charisma that attracts people to them. In fact, it's likely that most self-professed prophets would consider themselves light-bearers for sharing their unique insights with others, never for a minute stopping to consider what impact their teachings might be having or what repercussions they might portend if they were to be proven erroneous.

Moreover, not all the blame can be placed upon individuals. Sometimes entire organizations, groups, religious factions, or even nations perpetrate the most extraordinary nonsense—nonsense, by the way, that can be even more compelling because it may be based on supposedly ancient knowledge or sacred texts hundreds or thousands of years old. Dismissing the claims of a single individual is one thing, but it can be difficult to discount the long-held and deeply entrenched teachings of an entire institution such as the church, especially once these teachings become codified into the very fabric of that organization. In some belief systems, discounting any single teaching can be tantamount to rejecting the entire faith in toto, leading to excommunication and ostracization and, in some extreme cases, even death for heresy. It is difficult enough to abandon a belief system when there is no penalty for doing so; to reject one at the cost of one's family, livelihood, or very life is quite another matter, and often this is the main reason a person may hold on to an erroneous belief throughout a lifetime.

This latter point cannot be overemphasized. Peer pressure is a big element of doomsday beliefs—especially in the religious arena. By way of an example, for many fundamentalist Christians, not embracing the prevalent Second Coming mythologies made popular through the writings of Hal Lindsey, Tim LaHaye, Jack Van Impe, or any of the dozens of end-times preachers out there is often considered heresy. Some "rapture prophets" have even gone so far as to allude subtly to the possibility that some Christians may be "left behind" precisely because they don't anticipate the Second Coming, effectively punishing them not for rejecting Christ but for their lack of belief in a particular eschatological position.

It is a common human trait to want to be part of something larger than ourselves, and so we often seek to join a group of like-minded individuals to satisfy that desire. Such a mindset gives us a sense of security and makes us feel that we are an important and valuable part of the community, which is why some people are so often willing to embrace a supposed prophet's teachings *in spite* of the fact that they are neither proven nor compelling nor even particularly rational. And since the group often includes close friends—and sometimes even family—leaving it becomes synonymous with abandoning one's very roots and support system, making it necessary to suppress critical thinking skills in order to remain part of the "collective." Conformity is an integral part of such belief systems, especially if they are to gain converts.

Of course, this tendency to conform to groupthink doesn't apply only to doomsday cults, but to many religious, political, and social organizations as well. Those who predict social upheaval, environmental disaster, or technological catastrophe are no different from those who teach a literal Judgment Day; it is the same mentality that runs through all of them. Without this human characteristic, it would be extremely difficult to create "true believers," be they religious, political, or social in nature. Were people to approach their beliefs critically, cults and even many political action groups—along with most conspiracy theories—would largely disappear overnight, which really might hasten the start of a true golden age of humanity.

Chapter Three

Failed Doomsday Prophecies of the Past

It may end later, but I see no reason for its ending sooner. This I mention not to assert when the time of the end shall be, but to put a stop to the rash conjectures of fanciful men who are frequently predicting the time of the end, and by doing so bring the sacred prophecies into discredit as often as their predictions fail.

—Sir Isaac Newton

The date was October 22, 1844. Throughout much of New England, tens of thousands of men and women—some of them dressed in glistening white gowns in anticipation of their inevitable ascent to heaven—waited in breathless anticipation for Jesus Christ's imminent return. Followers of a respected New York farmer and self-taught Bible scholar named William Miller, all of them were confident their leader—despite the fact that twice before he had set dates that turned out to be wrong—was correct in his calculations, and that October 22, 1844, was *the* day. In fact, so sure where some of them that Christ was at that very moment preparing to return to Earth in glory and power that they had forgiven debts, sold their goods and possessions, and had even given away their savings to charities, secure in the knowledge that they would not need them.

Thousands of these "Millerites"—as they called themselves—waited patiently throughout the day for Christ to return, and as day gave way to dusk, an even greater sense of anticipation worked its way through the crowd. Their Lord, after all, had been placed in the tomb just as the sun was setting; was it possible he would return for them at the same time of day?

Eventually many of them found themselves upon hilltops sitting beneath the stars of a brilliant night sky, undeterred in their certainty that this time their prophet was right. Midnight came and a few wept or cried out in fear, expecting to be carried away at any moment. Others held tightly to their families while some sang and chanted, but most simply sat quietly and prayed silently to themselves as they looked for the first signs of their Lord's return.

But nothing happened. All was quiet on Earth, and as the clock tolled midnight without incident, a few of the congregants began to lose heart. Eventually the most discouraged started making their way back down from the hilltops, disappointment etched on their tired faces. By dawn the last of them had returned to their homes and to what remained of their lives, their anger at having been disappointed once more not only severely testing their faith but, in many cases, utterly destroying it.

Not all lost faith, however. A few continued to look daily for Christ's return for weeks and even months afterward, only to discover that each newly calculated date was also erroneous. One Millerite follower even taught that Christ had made it as far as our atmosphere and was even then sitting on a white cloud waiting to be prayed down, but again to no avail. No matter how hard they believed, prayed, calculated, or hoped, it seemed that Jesus was delaying his much anticipated Second Coming for some reason known, apparently, only to himself.[1] These few remained so unshakable in their faith, however, that they were to serve as the nucleus for what was to become the

---

1. Some managed to circumvent this, however, by claiming that Christ's return had genuinely occurred exactly as Miller had calculated, but that it was an "invisible" manifestation—whatever that means.

modern Seventh-day Adventist church, one of the largest and fastest growing Protestant denominations in the world today.

The majority, however, left the ranks of the church forever, many returning to their previous denominations or choosing never to visit a church again. For them and their disgraced leader—who was to die in obscurity just five short years later—their lives would never be the same, all because of a single man working under the presumption that he could predict that which even Jesus himself said that "no man could know," thereby serving as an object lesson to the many who would follow later.

Rise of the Date-Setters

But what happened? How could William Miller, by all accounts a sincere and devout farmer turned Baptist minister, have been so mistaken? He had gotten his date from a careful and meticulous study of the Bible itself, which, being the literal Word of God, *had* to be correct. So how could he have been so mistaken?

Of course, the Great Disappointment of 1844—as the event was later to be called—was not the first failed prophecy, nor would it be the last. Having considered what prophecy is and why we believe it, before exploring the 2012 prophecy craze and other prognostications of impending doom some anticipate occurring over the next few decades, it might be helpful to explore how predictive prophecies have fared in the past. Obviously, to thoroughly examine the entire history of failed prophecies would require a far larger volume than I'm prepared to offer here, but it might be useful to be aware of some of the more famous unrealized doomsday predictions throughout history so we might acquire a solid understanding of how they have emerged, developed, and evolved over the centuries.

I've started with the most famous of all failed American prophecies—the Great Disappointment of 1844—not because it was the most anticipated or most recent failed prophecy, but because it is so typical of those that do fail. The fact is that many churches have picked dates for Christ's return and each has followed a similar pattern of great disillusionment followed by a drastic decrease in membership, a slow rebound, and, eventually, another date being set, at which point the cycle starts over again. And the strange thing is that nothing is learned from these failed efforts; even today men and women, with great sincerity and conviction, continue to do the same thing, with the same zeal and earnestness that William Miller and his followers showed over a century and a half ago. It's not unlike a drug addiction in which a junkie wakes up in a hospital emergency room swearing to start anew only to end up overdosing again the very next day.

This fascination with dates—particularly having to do with "end-times events" (often commonly referred to as Judgment Day)—have been a part of nearly every faith system since antiquity. To get an idea of how prevalent they have been over the centuries, let's take a quick trip back through time to get a small taste of how successful our forebears' prognosticating turned out to be.

While doomsday prognostications have been a part of almost every culture, it is within Western faith traditions that they have been most prevalent. Probably the earliest doomsday prophecy known to most people is that of Noah and the ark, a story at least three thousand years old and, quite probably, far older.[2] Of course, as the book that contains the story—Genesis—was written long after the flood supposedly took place, it wasn't actually a prediction in the traditional sense of the word. However, insofar as Genesis tells us that God instructed Noah to build an ark because he was going to destroy the earth and all its inhabitants with a great flood, it suggests that Noah had been forewarned, making the account, in a way, a retelling of a prophetic proclamation. Unfortunately, being written after the fact makes it impossible to judge whether God's prediction of a great flood was accurate or not (and, further, it can be argued that God does not predict events but *initiates* them), but that's beside the point. What's important to recognize is that Noah's flood sets the stage for all

---

2. Most scholars today generally agree that Noah's account was a Semitic version of the Sumerian Epic of Gilgamesh, thought by many to be at least 4,200 years old and probably much older.

doomsday prophecies that were to follow, and continues to be the baseline event against which many doomsday prophecies are measured today.

Since then, both Judaism and Islam have been imbued with end-times themes, with Jewish expectations of a coming Messiah and anticipation among Muslims of the arrival of the Mahdi playing major roles in both of those faiths' traditions. In fact, Jewish expectations that the Messiah was to be revealed during the first century were rampant throughout Israel at the time and may have been the catalyst behind the Jewish revolt of 66–71 CE. Assuming that their revolt against their hated Roman occupiers was blessed by God and therefore was destined to be successful in the same way an earlier revolt had been successful against the Seleucid rulers two centuries earlier,[3] the Jews embarked on a path that practically ended in their own annihilation, all in an effort to realize the return of the Messiah. This is probably the first major instance in history in which an entire nation essentially committed national suicide in an effort to realize prophecy.

However, no religion emphasizes the end times more than Christianity, with its expectations of Christ's Second Coming and the Battle of Armageddon. Not remarkably, then, the majority of doomsday prophecies throughout history, at least until the last century or so, have been Christian-based. The reason for this is obvious: Christianity is uniquely based upon the belief in the physical resurrection and the equally physical return of Jesus of Nazareth to Earth. The "Second Coming" is the cornerstone of the faith, without which the entire belief system crumbles. In other words, without the physical return of Christ, which is considered by the church to be the single culminating event of history and, as such, synonymous with Judgment Day, there is no Christian story. As a result, Christians in the first century lived in constant expectation that Jesus would physically return during

---

3. In 167 BCE, Jews led by the Maccabean brothers successfully overthrew the Seleucid Dynasty, thereby winning a brief period of independence for Israel—a short-term victory that was to end a century later with the annexation of the country by the Romans in 63 BCE.

their lifetime to finish the work that had been cut short with his death, a major theme that is evident throughout the New Testament writings of both the disciples and the apostle Paul.

However, despite the stubborn refusal of Christ to return in bodily form during the first few decades of the church, there were those who continued to look for Christ's imminent return long after all of the disciples had passed from the scene. Interestingly, though, most doomsday prophecies have not come from the established churches themselves—which tend toward caution where the subject is concerned—but largely from the laity or from smaller splinter groups within the church (the aforementioned Millerites being a good example). Furthermore, only in the past couple of centuries has the church at large formally paid much attention to end-times scenarios in general, with the bulk of the enthusiasm for such ideas coming from fundamentalist denominations within Protestantism, particularly Baptists and other evangelical strains.

The Anticipation of Christ's Return

Perhaps one of the earliest and most influential "end-times themed" groups to come along in the aftermath of Christ's unrealized bodily return in the first century were the Montanists, who were followers of the tongues-speaking[4] prophet Montanus and his two female followers, Priscilla and Maximilla. It was Montanus who, in 156 CE, taught that Christ would come again within the lifetimes of people then living to establish a new Jerusalem at Pepuza, in the land of Phrygia. Despite the failure of Jesus to return, the cult lasted for several centuries and counted several noted Christian leaders among its followers.

---

4. Speaking in tongues—known technically as *glossolalia*—is the supposed ability to speak coherently in a language not known to the speaker. In religious circles, it is thought this ability is generated by God via the Holy Spirit, making it a type of holy language that is occasionally prophetic in nature.

Another group of fairly famous doomsday proponents who flourished in the fourth and fifth centuries were the Donatists (named for the Berber Christian Donatus Magnus), who boasted a fairly large cult of Christians that looked forward to the world ending in the year 380. They disappeared shortly afterward, however, only to be replaced by followers of another theologian, Sextus Julius Africanus (c. 160–c. 240 CE), who claimed that the world would end six thousand years after the Creation, which, based upon his assumption that 5531 years had elapsed between the Creation and the Resurrection, would place doomsday no later than 500 CE.

So persuasive was his reasoning, it appears, that the important church fathers Hippolytus and Irenaeus accepted the figure, and were able to bring even more adherents into the fold long after Africanus was gone. Fortunately for them, they all were long dead by the year 500, thereby sparing them public humiliation for their presumptuousness. However, even in death Africanus was clever: just before his death he himself, probably after reconsidering just when the Creation might have taken place, went on to update his end-times date (which is something rarely seen until *after* a date has been missed) to 800 CE, thereby giving himself a three-hundred-year cushion. In any case, by the time the later date arrived, Africanus' prognostications had been largely forgotten or overlooked in the church's preoccupation with maintaining its hold on power in the aftermath of Rome's collapse as an empire.

By the seventh century CE, doomsday prophecies had taken a back seat to more immediate concerns, most particularly the growing battle with the forces of a new religion out of Arabia known as Islam. This new faith, based in part on Judaism and Christianity with just a dash of indigenous Arab beliefs thrown in, not only swept the church from the holy land but had seized all of Asia Minor and Africa and threatened to overwhelm Europe as well. As far as the church was concerned, the threat posed by Islam *was* doomsday, and though Christ's return and with it Judgment Day remained a major tenet of the faith, end-times teachings were not particularly popular for the next few centuries.

By the end of the first millennium, however, with the immediate threat posed by Islam temporarily met and the thousandth-year anniversary of Christ's birth looming, end-times scenarios once more became popular. While there is some debate among historians about how significantly doomsday beliefs affected the populace at the time, it was probably substantial—if the turning of the most recent millennium is any indicator, and considering the even more pervasive superstition of the times.[5] Considering the role the number 1,000 often plays in Scripture, it would be hard to imagine that the turn of that millennium would not be considered extremely significant, especially among those who took the Book of Revelation's thousand-year-millennium reign of Christ literally.

After the change-of-millennium "danger period" passed quietly, however, it appears that end-times predictions once more fell out of favor to a large degree, making the eleventh through fifteen centuries relatively devoid of such prognostications. However, there were a few who kept the doomsday fires burning. Perhaps the best known of these was the Italian mystic Joachim of Fiore (c. 1135–1202), who declared that, at least according to his calculations, the world would end sometime between 1200 and 1260 CE. After 1260 passed uneventfully, how-

---

5. While early-twentieth-century scholars largely dismissed the notion that the change of the millennium created any great disturbance among the population of Europe, others have since challenged this premise, suggesting that there is far more evidence for apocalypticism being rampant around the year 1000 than had been previously thought. Some also note that over the last half-century medievalists have come to view the period around the turn of the millennium as a time of great social and cultural transformation, making it possible to speculate that doomsday expectations around this period may have been more influential than earlier historians were willing to concede. Furthermore, evidence suggesting that doomsday fever occurred around the turn of the millennium and that it continued right up to 1033 CE—the thousandth anniversary of Christ's death and resurrection—is contained within the writings of the Burgundian monk Rodulfus Glaber (985–1047), as well as other chroniclers of the age.

ever, his followers stepped in and revised the dead mystic's doomsday date to, in turn, 1290, 1335, and, finally, to 1378. (Sound familiar?)

However, things really got rolling in the sixteenth century, when no fewer than eighteen dates—and probably many more—were suggested by various clergymen, mystics, and astrologers throughout the century, many of which caused considerable panic amongst the populace. Perhaps the most noteworthy of these failed prophecies came on February 1, 1524, when 20,000 Londoners abandoned their homes after various English astrologers calculated that a great flood in the city was imminent. There is no historical record that tells us how these same English astrologers fared after the date came and went without incident, however.

But the sixteenth century was just a harbinger of things to come. With the advent of the printing press, apocalyptic writings became, relatively speaking, more and more accessible to both the clergy and an increasingly literate public, resulting in a burgeoning of date-setting as many theologians, mystics, and astrologers took turns trying to read the signs in an effort to determine when the end would come. Much like today, almost every year was picked by someone as the most likely date of doomsday, with some dates garnering much more interest—or fear—than others. One date that especially stood out was 1666, which was the combination of a millennium (1000) and the mark of the beast of Revelation (666). While there was no global doomsday that year, 1666 was the year a great fire gutted London, killing hundreds and destroying 70,000 of the 80,000 homes in the city, making the year a kind of doomsday for Londoners. For the rest of the world, however, the date passed uneventfully, and soon 1666 was consigned to the dustbin of history.

Lest anyone imagine that such soothsaying was limited only to occultists and overzealous religious types, it should be noted that Scottish scientist and mathematician John Napier determined, based upon his study of the Book of Revelation, that doomsday would occur in 1688. Even the famous Sir Isaac Newton was heavily influenced by the writings of the English scholar Joseph Mede, who set 1660 as the date for things to wrap up on planet Earth. For the most part, however, churchmen were doing most of the prognosticating, with Deacon William Aspinwall, the leader of the Fifth Monarchy movement, which was dedicated to installing a theocracy on Earth, picking 1673 as *the* date, and Anglican Rector John Mason choosing 1694 as the year it would all come to a fiery end. Even the notorious witch-hunter Cotton Mather got into the act, selecting, in turn, 1697, 1716, and 1737 as the year Christ would make his long-anticipated return. Fortunately, it appears these people did little appreciable damage to society as a whole with their failed soothsaying, probably because their teachings were so localized.

Prophecies of the Nineteenth and Twentieth Centuries

Up to that point, most doomsday groups were regionally limited and/or had comparatively small followings, but all that began changing in the nineteenth century with the advent of the Millerite movement, mentioned in the opening of this chapter. For those unfamiliar with the man, William Miller (1782–1849) was a New York farmer turned Baptist minister who made studying Scripture his life's passion. Embarking on a careful study of the Old Testament—especially the book of Daniel and its obtuse use of various numbers—during the 1830s, he finally came to the conclusion that Jesus Christ would return on October 22, 1844.[6] How he arrived at that precise date is the result of a fairly complex series of calculations, but suffice it to say that by 1840 his powers of persuasion were sufficient to induce upward of 50,000 (some estimates are as high as 500,000!) of his fellow New Englanders to buy off on his teachings. When the day came and went without Christ's return, however, the disappointment was, to put it mildly, more than a little palpable. Miller lived out the final years of his life a virtual recluse, devastated by his great disappointment but never for a

---

6. Although Miller was forced to "refine" his date twice, finally settling on October 22, 1844, as *the* date, that didn't dissuade his followers from taking him at his word.

moment giving up on his belief that the Second Coming was "imminent."

Despite Miller's error, however, his teachings appear to have had a considerable impact upon other churches of the period. How much influence his teachings may have had is difficult to gauge, but by some accounts it was considerable. For example, it's likely his ideas influenced the teachings of another contemporary of his, Joseph Smith, Jr., founder of the Church of Jesus Christ of Latter-day Saints (a.k.a. the Mormons), who wrote in 1835 that a heavenly voice told him that "if thou livest until thou art eighty-five years old [i.e., December 1890], thou shalt see the face of the Son of Man [Christ]."[7] This was not a prophecy Smith was to see fulfilled, however, as he died in a Carthage, Illinois, jail shootout in 1844.

However, Miller's teachings appeared to affect no single religious organization in America more than the Jehovah's Witnesses—a group founded in 1874 by a onetime Millerite and Congregationalist by the name of Charles Taze Russell (1852–1916). Since its inception, no denomination has been more guilty of repeatedly setting dates than have the JWs, usually to their own detriment. Though he proposed several dates for Christ's return starting in 1874, Russell's most famous doomsday date was October 1, 1914, a date that happened to dovetail nicely with the start of the First World War (which Russell—an ardent pacifist—and his followers believed to be the start of the Battle of Armageddon). When Christ didn't physically return on schedule, though, Russell—taking a cue from some of Miller's followers—simply suggested that the Lord had returned "invisibly" instead, though without defining exactly what that meant. In any case, it didn't seem to impair the vigor of the church to any great degree, which continued to see extraordinary growth in the intervening decades after Russell's death in 1916.

---

7. *The Doctrines and Covenants of the Church of Jesus Christ of Latter-Day Saints, Containing Revelations Given to Joseph Smith, The Prophet*, 130:15. Salt Lake City: The Church of Jesus Christ of Latter-Day Saints, 1921, p. 238.

Undeterred by their 1914 "miss" (or, apparently, unaware of Christ's aforementioned "invisible" return), church leaders, under the leadership of Russell's energetic successor, Joseph "Judge" Rutherford, went on to name several other years as the date for the Savior's return. The years 1918, 1920, 1925, and 1941 were all proposed at one time or another, but each passed uneventfully (with the exception of 1941, when World War II raged around the world). These "misses," however, seemed to have had little appreciable deleterious effect on the church or its continued growth for several decades.

However, the church's penchant for date-setting almost did it in when it announced that 1975 was to be the year of Christ's final, visible return. (This sure-fire date was based on the belief that Adam was created in the year 4026 BCE, making 1975 the 6,000th anniversary of that miracle.) The church encouraged Jehovah's Witnesses to sell their homes, quit their jobs, and forego all planning for the future in deference to praying and doing door-to-door evangelizing until the end came. With the dawn of 1976 came considerable buyer's remorse, and with it a general exodus from which it was to take the church decades to recover. As a result, the organization has been considerably more careful about date-setting ever since, although apocalyptic, end-times beliefs continue to permeate their teachings to this day.

Another self-professed Bible scholar who saw a similar decline in membership after failing to deliver on an end-times prophecy was Herbert W. Armstrong, of the Worldwide Church of God. Armstrong was quoted in a 1972 *Atlantic Monthly* interview as predicting the "beginning of the end" starting in January of that year. Just as many Jehovah's Witnesses experienced in 1975, many of his members, then numbering in the hundreds of thousands, suffered great hardship, as they had given most of their assets to the church in the expectation of going to Petra (Armstrong's euphemism for heaven), where such worldly possessions would be useless. The failure of this prophetic scenario to take place according to plan may have been one of the reasons why church membership began to fall off so prodigiously afterward.

Date-setting was not confined to Christian-based churches, however. New Age groups also contributed to the end-times debate, with numerous gurus and practitioners of all ilks setting various dates for the end. No group, however, got as much attention as did the founder and spiritual head of the Church Universal and Triumphant, Elizabeth Clare Prophet (a.k.a. Guru Ma), who taught that a nuclear war would start on April 23, 1990, and convinced many of her followers to stockpile food and guns in underground bomb shelters in Montana in response. Such preparations proved to be a bit too much for the federal government, however, which promptly seized her arsenal and convicted a handful of her followers—including her husband—on weapons charges. Curiously, the lack of nuclear war did not immediately spell the end for her movement, as her followers continued to hunker down in their Montana bunkers for some time afterward.

Lest anyone imagine that end-times scenarios were confined to faith-based groups, science—or, should we say, pseudoscience—also got in the act, starting with the writings of one Piazzi Smyth (1819–1900). A past astronomer royal of Scotland, in 1860 he wrote a book titled *Our Inheritance in the Great Pyramid*, which spread the belief that secrets are hidden in the dimensions of the great pyramids of Egypt. Smyth claimed, and presumably believed, that the dimensions of the pyramid (which he measured in inches) were a God-given measure handed down through the centuries from the time of Israel, and that the architects of the pyramid could only have been directed by the hand of God. To support this contention, Smyth, after carefully measuring the pyramid by hand, noted that the number of inches in the perimeter of the base equaled one thousand times the number of days in a year, and found a numeric relationship between the height of the pyramid in inches to the distance from Earth to the sun, measured in statute miles. He also advanced the theory that the Great Pyramid was a repository of prophecies that could be revealed by detailed measurements of the structure.

Somehow he concluded from all this that doomsday would start in 1882, a date he was forced to reconsider as New Year's Day of 1883 dawned. Undaunted, the determined Mr. Smyth recalculated the date to 1892 and, when that year also passed without incident, recalculated again to 1911. All was not lost, however; although all his guesses proved to be conspicuously incorrect, his exhaustive work on the Great Pyramids did prove influential for a time and even served as inspiration to some of Charles Taze Russell's (of Jehovah's Witnesses fame) end-times calculations.

Since Piazzi Smyth's day, science's role in manufacturing doomsday predictions has grown rapidly, with various environmental and nuclear-war alarmists rising to warn the world of impending catastrophe. From meteorologist Albert Porta, who predicted in 1919 that the conjunction of six planets would generate a magnetic current that would cause the sun to explode, to NASA engineer Edgar Whisenant (1932–2001), who self-published a book called *88 Reasons Why the Rapture Will Occur in 1988* (a book that sold four million copies in a few short months), science has always been a player in the doomsday game.

Yet none of these efforts gained as much attention as did the 1974 bestseller *The Jupiter Effect*. Authored by two professional astrophysicists, John Gribbin and Stephen Plagemann, the book premised that a rare alignment of all nine planets in 1982 would create a combined gravitational pull that would place huge stresses on the planet's tectonic plates, causing killer earthquakes and severe changes in the earth's climate. The book was supposedly intended simply as an exercise in astrophysical speculation, but it was subsequently misinterpreted and abused by end-of-the-Earth enthusiasts, and once 1982 passed without incident, it caused considerable embarrassment and professional damage to the two men's careers.

The book did have an impact on the debate, however, for the idea behind the book—that it was nature and not the gods one needed to be worried about—served as the backdrop for numerous "death from space" scenarios that have come along since. We'll examine some of these real-life dangers—from rogue asteroids to exploding comets—in more detail later, but suffice it to say that not all end-times scenarios are the product of fertile religious imaginations; scientifically inclined individuals are just as capable of creating their own doomsday scenarios, given enough time and a universe of "what-if" possibilities to consider.

Conclusions

Certainly the history of soothsaying is not a particularly impressive one. In fact, one would be hard-pressed to give those predictions of the past even a 1 percent success rate (and, even then, what constitutes success would be open to interpretation).

Of course, there are those who would claim that while the ancients might have been wrong more often than right, there were a few men and women who might really have been correct in their prognostications, and only have yet to realize them. After all, it is argued, even if we can make the case that doomsday predictions of the past have proved to be uniformly dismal failures, that doesn't prove that *all* such predictions are necessarily nonsense. That's like saying that because nearly everyone who plays the lottery loses, it is impossible for *anyone* to pick the correct numbers—a prospect demonstrated to be erroneous with each successful drawing.

And the two "prophets" most commonly looked upon as being genuine seers today by millions of people around the world—even if their prophecies have yet to be realized—are Michel de Nostredame and a humble little man from Hopkinsville, Kentucky, named Edgar Cayce, both of whom we will examine in some detail in an effort to ascertain whether they truly were men who genuinely had a gift for prophecy or whether they are just another product of our collective desire to believe.

Chapter Four

Nostradamus: The Seer of St. Rémy

Although he was little known outside of the courts of France during his lifetime, since his death almost 450 years ago Nostradamus has become so synonymous with the title of prophet that today some consider him to be the benchmark by which all other such seers are to be judged. Since a complete treatment of the man and his prophecies could easily take up the majority of this work and sidetrack us from our main premise, this will be, by necessity, a very abbreviated examination of the man and the hundreds of prophetic quatrains he penned throughout the course of his life. Partially this is for the sake of brevity, but the main reason is that I want to deal specifically with his end-times prophecies, which constitute only a tiny fraction of the many prophecies he penned and so render the bulk of his work irrelevant to our discussion. However, for those whose knowledge of the man and his story is limited, a greatly condensed biography as well as a brief discussion of a few of his more famous prophetic quatrains may be in order.

Born in Saint-Rémy-de-Provence in the south of France in December of 1503, Michel de Nostredame—better known as Nostradamus—had by all accounts what could only be described as a most unorthodox childhood. Born to Jewish parents who had converted to Roman Catholicism (apparently in an effort to avoid persecution by the local church), young Nostradamus grew up in a family that managed to blend together both Catholic and Jewish customs—along with just a hint of mysticism—into a unique mix that couldn't have helped but make the young man's view of the world different from many of his contemporaries.

At the age of fifteen, young Nostradamus entered the University of Avignon to study liberal arts but was forced to leave when the university closed in the face of an outbreak of the plague. After leaving Avignon, Nostradamus (according to his own account) traveled the countryside for eight years, researching herbal remedies until in 1529 he entered the University of Montpellier to study medicine. Eventually expelled when it was discovered that he had been an apothecary (the medieval equivalent of a pharmacist and a trade that was banned by university statutes), Nostradamus continued working as an apothecary while he taught himself as much as he could about astrology. So knowledgeable was he to become with regard to the "dark arts" that he eventually became one of the most celebrated astrologers and seers in the courts of France—a title he proudly held right up to the time of his death in 1566.

If that had been all he did in life, most would never have heard of the man (indeed, he was little known outside of France until the twentieth century). What kept his name alive was the fact that he was an astrologer who was a most prolific writer as well and managed to record the bulk of his prophetic utterances in a book entitled *Les Propheties de M. Michel Nostradamus* (generically referred to simply as the *Centuries*), the first edition of which appeared in 1555 and has been in print ever since—a feat matched only by the Bible.[1]

In a poetic style reminiscent of limericks, Nostradamus wrote his predictions in the form of quatrains—a type of rhyming four-line poem or stanza common in European literature of the time. In all, Nostradamus penned a total of 942 such quatrains over a five-year pe-

---

1. He also penned a popular series of *almanachs* (detailed predictions), and *prognostications* or *présages* (more generalized predictions), though his *Centuries* remain the most popular of all his writings.

riod, which he organized into *Centuries*: nine groups of one hundred quatrains each (with one *Century* having only forty-two quatrains). They were written in Old French, an archaic predecessor of modern French, as well as Latin, Greek, and Italian, making their precise translation into English difficult. They were also full of esoteric metaphors and anagrams that included few dates or specific geographical references, which make their meaning even more enigmatic. And, to top it off, they weren't even arranged in chronological order, further enhancing the difficulty of building a timeline around Nostradamus' numerous quatrains.[2]

Moreover, not everything Nostradamus wrote was original. It's been demonstrated that he copied from numerous classical historians and other sources in compiling his writings, as well as borrowed heavily from medieval chroniclers such as Villehardouin and Froissart. In fact, much of his prophetic work appears to have been paraphrased from other ancient prophetic collections (most of them Bible-based), supplemented with references to historical events and anthologies, which he then projected into the future with the aid of horoscopes. This explains why some of his predictions mention such long-dead figures as the Roman general Sulla and the emperor Nero. Furthermore, it appears that one of his major prophetic sources was evidently the *Mirabilis Liber* of 1522, which contained a range of prophecies by Pseudo-Methodius, Joachim of Fiore, Savonarola, and others. While at first glance this would seem to make him more of a plagiarist than a prophet, it should be noted that modern views of plagiarism did not apply in the sixteenth century, allowing authors to frequently copy and paraphrase passages at will without acknowledgement, especially from the classics.

---

2. How Nostradamus created his quatrains is an interesting story in itself. By most accounts, he would stare intently into a bowl of water until he could make out some sort of a "vision" in the swirling ripples, which he would then write down. Undoubtedly he may also have made use of astrological charts as well, which he likely consulted in preparation for "reading the waters."

However—and despite these difficulties—interest in Nostradamus' work remains high in our popular culture, with some of his prophecies even being assimilated in an effort to authenticate the so-called Bible Code as well as other purported prophetic works.

So how accurate were his prophecies? While Nostradamus is often credited by many to have been uncannily accurate in his predictions concerning the future (from his perspective), just how accurate his predictions really were remain a source of considerable debate. The first problem is that it's unclear exactly which era the man is referring to in many of his quatrains. Most appear to be veiled references to political and historical events from his own time, while others seem to refer to the distant past or, possibly, to the far-flung future. Most, however, are written in such a vague manner that they can be made to speak to almost any major historical period in history. By way of an example, consider this quatrain from *Century* II:

> *Beasts ferocious from hunger will swim across rivers:*
> *The greater part of the region will be against the Hister,*
> *The great one will cause it to be dragged in an iron cage,*
> *When the German child will observe nothing.* (quatrain 24)

The reference to the "German child" and "Hister" is interpreted as a reference to Adolf Hitler by most Nostradamus buffs, due in large part to the fact that Hitler and *Hister* differ by only a few letters, which would seem to make it a futuristic prophecy from Nostradamus' perspective. However, that is an assumption with little to support it. The reference to *Hister*, for example, is not referring to a person but to a place—the Hister being the ancient name for the Danube River.[3] Additionally, the quatrain does not tell us what it means when it says " . . . the great one will cause it to be dragged in an iron cage / When the German child will observe noth-

---

3. Proponents correctly point out, however, that Hitler did indeed grow up on the banks of the Danube River, potentially giving the prophecy a double meaning. It's also interesting to note that Hitler himself thought of the prophecy as referring to him—though he interpreted it as denoting victory—and even used it as a propaganda tool.

ing." Certainly it is difficult to see how this would apply to Hitler or the Second World War, making the entire quatrain entirely too ambiguous to be useful.

Death of a King

However, it would seem that for Nostradamus to have achieved so exalted a status as a gifted seer he would need to have some obvious fulfilled prophecies to his credit. So are there any that are irrefutably accurate? Fans of Nostradamus point out one quatrain that does appear to be deadly accurate. In *Century* I, quatrain 35, the prophet writes:

> *The young lion will overcome the older one,*
> *in a field of combat in single fight:*
> *He will pierce his eyes in their golden cage;*
> *two wounds in one, then he dies a cruel death.*

This quatrain is generally considered to be a prediction of the death of French king Henri II, who died from injuries received in a jousting duel in 1559. In fact, just as the quatrain says, the king died when his opponent's lance pierced the king's headgear (the "golden cage"), shattered into fragments, and penetrated his right eye. He died eleven days later in terrible agony, apparently demonstrating that Nostradamus was indeed capable of occasionally getting "hits," to use prophetic parlance.

However, not everyone is convinced. The quatrain never mentions who the "older one" is or a jousting tournament, only a "field of combat in single fight." While this could be interpreted as meaning a jousting field, it's hard to see how it was combat; the king was not engaged in a battle for supremacy against "the young lion" but merely engaging in a friendly joust as part of a three-day wedding celebration for his recently married daughter. It's also not clear what is meant by the injured party having a "golden cage," as jousting helmets were never made of gold (though this could possibly be a metaphorical reference to his kingship). In any case, Henri II was known for his love of engaging in such dangerous sports and had been repeatedly warned by his advisors not to engage in jousting matches. Nostradamus, who was after all middle-aged at the time of King Henri's reign, knew this, thereby possibly making his prediction—as vague as it was—simply a good guess. It's really no different from predicting that a test pilot who enjoys pushing the envelope with each aircraft he flies will one day end up being killed by one of his test planes. Additionally, had this event not occurred, it would simply have become but one more of the man's many obscure prophecies rather than his crowning prophetic achievement.

Louis Pasteur?

However, such ambiguities do not deter Nostradamus' most enthusiastic proponents, who are quick to point to another of his many quatrains as evidence of his prophetic skills. This one is contained in *Century* I:

> *The lost thing is discovered, hidden for many centuries.*
> *Pasteur will be celebrated almost as a God-like figure.*
> *This is when the moon completes her great cycle,*
> *but by other rumors he shall be dishonored.* (quatrain 25)

Supposedly this is a reference to Louis Pasteur (1822–95), the French chemist and microbiologist best known for his breakthroughs in the study of and prevention of disease as well as for his work in germ theory that revolutionized modern medicine. If correct, this certainly would be an extraordinary feat of prognosticating, especially considering that it refers by name to a man who would not appear until more than two centuries after Nostradamus wrote the quatrain.

However, it's not as simple as that. The biggest problem with the quatrain is that, in the original French, the word *pasteur* can be translated as either "pastor"—a religious leader—or simply as "shepherd" (this, again, is where translating from one language, ancient French, into modern English causes problems). While the quatrain does contain the word *pasteur*, that may be better evidence of random coincidence (combined with the fact that Pasteur is not a particularly uncommon French surname) than a prophetic "hit."

As for the wording that the "lost thing" that is discovered after having been "hidden for many centuries" is a reference to bacteria—as some of his proponents maintain—this too is presumptuous. It could mean anything the reader wants it to mean, and besides, the wording seems to imply not that something new would be discovered, but that something that was previously known but had been subsequently lost would be rediscovered. As such, that would seem to eliminate bacteria as the source of this statement, since the existence of germs was unknown prior to the nineteenth century. Finally, it could be argued whether Pasteur ever achieved "God-like" status as a scientist and, of course, while he had his critics, it's hard to see how he was dishonored by rumors. In reality, he was actually hailed as a great scientist at the time of his death and remains considered as such to this day. Also, the moon completing her "great cycle" is normally understood to mean a lunar month (about twenty-eight days), but how does that figure into Pasteur's life? Obviously, the quatrain raises more questions than it answers.

September 11th Quatrains?

However, it isn't those quatrains that point to the past that capture most people's attention, but rather those that seem to refer to contemporary events or, potentially, near-future events (from our perspective). Probably the most-frequently-referred-to quatrain that seems to speak to modern events is one that supposedly makes reference to the terrorist attacks on New York and Washington DC on September 11, 2001. According to *Century* I, quatrain 87:

> *Earthshaking fire from the center of the earth*
> *will cause tremors around the New City.*
> *Two great rocks will war for a long time,*
> *then Arethusa will redden a new river.*

The key here is the term *New City*, which many take to mean New York City. That is a huge assumption, however, as *New City* could be a reference to any number of cities on the planet (one good prospect being Naples, Italy, which was often referred to in antiquity as the "new city"). Additionally, it is difficult to see how the collapse of the World Trade Center towers could be thought of as causing "earth-shaking fire" and "tremors" except in a metaphorical sense. Furthermore, while the "two great rocks" that will "war for a long time" could describe the ongoing struggle between the west and militant Islam, again such an interpretation is a most precarious one. In fact, the case could be made that the quatrain appears to be pointing toward a major geological event rather than some sort of terrorist attack, especially if one recognizes Naples' (the New City?) proximity to Mount Vesuvius and its propensity toward seismic activity—a fact well known in Nostradamus' day.[4] In that, then, his prediction wouldn't be much different from predicting an earthquake in San Francisco or Los Angeles. Without tying the prophecy to a specific date, such a prediction would be a pretty safe bet.

But that's not the only quatrain Nostradamus' supporters point to as being suggestive of the September 11th attacks. Another of his nearly 1,000 quatrains that has been interpreted by some as being a reference to the WTC collapse comes from *Century* VI, quatrain 97, which reads:

> *At forty-five degrees the sky will burn,*
> *Fire to approach the great new city:*
> *In an instant a great scattered flame will leap up,*
> *When one will want to demand proof of the Normans.*

Again, however, this "prophecy" is not very clear. Certainly, if the "great new city" it's referring to is New York City, Nostradamus missed it by three hundred miles (it lies at about 40 degrees north latitude, not 45), and it's especially difficult to see precisely how the Normans figure into the mix.

---

4. In fact, Naples has been rocked by earthquakes twice since Nostradamus made his prediction, once in 1693 and again in 1980. Could the "two great rocks" to which he refers even be metaphors for strata layers being divided by a fault line?

What's interesting about the 2001 attack, however, is how so many of Nostradamus' quatrains have been distorted or fabricated in an effort to make them "work" as a 9/11 prophecy. One notable example of this that has been making the rounds since 2001:

> *In the City of God there will be a great thunder,*
> *Two brothers torn apart by Chaos,*
> *while the fortress endures, the great leader will succumb.*
> *The third big war will begin when the big city is burning.*

While impressive—especially the part about the third big war starting when the big city is burning—it turns out that *this quatrain is a complete fabrication.* Despite that, however, it remains included as authentic on many websites and has even made its way into publication, demonstrating how easy it is to both fabricate and promulgate bogus prophecies. This has had the effect of making it even more difficult to determine whether there is any validity to Nostradamus' claims of being a seer of remarkable skills, and further pollutes the doomsday environment with ever-increasing amounts of garbage.

Nostradamus' *Real* End-Times Quatrains

But did Nostradamus predict a third world war or doomsday, as many of his proponents believe? Actually, Nostradamus never sets a date for doomsday in any of his writings. The closest he comes is in a letter to a friend, in which he states that his predictions cover thousands of years, out to the year 3797. However, this was not a prediction that the world would end at that point, but simply a statement about how far his projections would extend. In fact, if looked at from another perspective, the fact that he says his predictions extend that far into the future suggests that the many doomsday prophecies made up to now are all going to be wrong, for if they aren't, there would be no 3797 to predict through.

However, there are a few quatrains that appear to be tantalizingly close to setting dates. Probably the most famous of his quatrains that does this is found in *Century* X, quatrain 72, which reads:

> *When 1999 is seven months o'er*
> *Shall Heaven's great Ruler, anxious to appease,*
> *Stir up the Mongol-Lombard king once more*
> *And war reign haply where it once did cease.*[5]

While it's interesting that Nostradamus specifically mentions July of 1999, the problem is that it's difficult to find anything of great historic import that occurred at the time. There was a meeting of world leaders in the war-torn city of Sarajevo about that time trying to hammer out a peace deal over the future of the Balkans, but what that might have to do with the rest of the verse—not to mention who "Heaven's great ruler" and the "Mongol-Lombard king" were—is anybody's guess. As such, a popular quatrain that garnered much attention prior to 1999 turns out to be as big a bust as the vast majority of Nostradamus' other prophecies.

There are a few quatrains, however, that his proponents claim predict the appearance of three antichrists: Napoleon, Hitler, and a third yet to be revealed; and because the emergence of the antichrist (the third, or yet to be realized one) has traditionally been tied to the end of the world, that would make such a prophecy—if indeed it exists—apocryphal in nature. The most famous of these is found in *Century II*, quatrain 62, where Nostradamus writes:

> *"Mabus" then will soon die, there will come*
> *Of people and beasts a horrible rout:*
> *Then suddenly one will see vengeance,*
> *Hundred, hand, thirst, hunger when the comet will run.*

Unfortunately, it's not at all clear who "Mabus" is, or whether it's even the name of a person at all. This doesn't stop many of Nostradamus' more imaginative proponents from deciding it's some sort of

---

5. Another translation shows this as *The year 1999, seventh month, From the sky will come a great King of Terror: To bring back to life the great King of the Mongols, Before and after Mars to reign by good luck*, which demonstrates how even the various translations can disagree.

arcane reference to a future antichrist, while others suggested it was a veiled reference to Saddam Hussein (*mabus* spelled backward being *subam*, which, if you turn the *b* into a *d* and the *u* into an *a*—oh, and add a second *d*—would spell *saddam*), while still others creatively merged together the names of George W. Bush and Osama bin Laden to get OsaMABUSh! Clearly, given the proclivity to find hidden meaning in even the most obtuse text and armed with a creative imagination, Nostradamus' vague writings can be made to speak to just about any person, political situation, or historic event one desires, and to do so with all the reliability of a fortune cookie.

Conclusions

Aside from the clumsy attempts of amateur hoaxers and the determined efforts of Nostradamus' most ardent proponents to tie even the vaguest examples of his prognostications to actual historical events, we still need to ask whether the man may have had legitimate prophetic powers. While this is purely an opinion, the evidence points to "no." Nostradamus, for all his sincerity, was probably a well-intentioned man who truly believed he had a knack for prophecy, but more likely was deceiving himself (and countless millions ever since). It's also possible that many of his quatrains were not intended to be taken as prophetic allusions at all but as a type of political commentary on the events of his time—couched, however, in obscure and metaphorical language designed to protect himself from those who might wish him harm if they were to understand a particular quatrain's true intent (or target). Such would not only have been prudent—especially if he wished to continue to enjoy the lavish and privileged lifestyle to which he had become accustomed—but absolutely essential if he wanted to keep his head.

In the end, it is becoming increasingly evident that Nostradamus' prophecies have been kept alive only through the efforts of his most passionate supporters, without whom his writings would probably have long ago been forgotten. That is both a blessing and a curse: a blessing in that such attempts to keep his words alive have given us some insight into the world of sixteenth-century France, a world as distant and fascinating to us today as ours would surely have been to Nostradamus in his age, and a curse in that they give the world more end-times nonsense to debate over. If nothing else, we should be grateful that the man's writings serve as a time capsule that allows us to peek into the intrigues and mindset of an era lost to us in history.

Chapter Five

Edgar Cayce: The Sleeping Prophet

Of course, Nostradamus is not the only voice who has made bold predictions of end-times events, nor is France the only country so endowed with such prophets. America has its own modern Nostradamus, though he was a very different type of person from his sixteenth-century French counterpart. However, he is as well known for his predictions today as Nostradamus was known in his time, and he is considered by many to have a considerably better record of success than any prophet of the modern age. So who is this gifted person?

His name is Edgar Cayce—also known as the sleeping prophet.

For those unfamiliar with the name, a brief biography may be in order. Edgar Cayce was an unlikely candidate for a man destined to be one of the most successful prophets in modern history. In fact, his life started out about as undistinguished and unspectacular as one could imagine. Born on a small farm near Hopkinsville, Kentucky, in 1877, Cayce proved to be an average but reasonably intelligent boy, who, due to the demands of working a small family farm, was forced to quit school early, resulting in him never receiving more than an eighth-grade education. While this was not unusual in the farmlands of the nineteenth century, it did mean that young Edgar would have to make his way in the world with few prospects and, by most estimates, little chance of escaping the poverty of his youth.

One thing different about young Cayce, however, was that even as a boy he appeared to exhibit unusual physic abilities, including, at least according to later biographers, an ability to speak with spiritual entities (i.e., angels) as well as to the dead (among them the ghost of his late grandfather). Perhaps most remarkable of all, though, was a supposed ability to retain all the information in a book by simply sleeping with the book under his head. Whether this ability was ever proven is difficult to ascertain, as most stories regarding the young Cayce's life are largely anecdotal, but it was apparent by almost all accounts that this simple and unassuming man had an interesting childhood that was to serve him well in later life.

Having little education forced young Cayce to take what work he could find to support himself. As such, he worked briefly at a dry-goods store and a bookstore before going into the insurance business with his father. It was while working as an insurance agent in 1900 that his life took an unfortunate turn when he developed a severe case of laryngitis—a condition that not only made it increasingly difficult to sell insurance, but one that refused to respond to traditional treatments of the day. The condition proved so severe and apparently untreatable that it eventually forced young Cayce to give up on the insurance business, which made him understandably depressed and anxious about how he would make a living. However, he eventually found employment working as an apprentice in a photography studio in Hopkinsville, which was a job that put less strain on his vocal cords.[6] Unknown to him at the time, however, his illness was not the curse he assumed it to be, but would prove to be the single event that would profoundly change his life.

One night, a traveling hypnotist arrived in town and, hearing of young Cayce's predicament, offered to try and cure his laryngitis through hypnotism. Now hypnotism (or mesmerism, as it was known in those days) was a fairly new phenomenon and so Cayce was a little

---

6. His skills as a photographer were also to serve him as a secondary source of employment throughout his life.

hesitant to take the man up on his offer—especially in front of a crowd. But he was growing desperate enough to try anything and so consented to his offer. Fortunately, Cayce proved to be an especially good subject and was soon in a deep trance, at which point he was able to speak in a normal voice for the first time in months. When he was brought out of the trance state, however, he found he was once again unable to speak, a phenomenon that repeated itself each time the hypnotist put him under.

Apparently unable to effect a permanent cure for Cayce's condition, the hypnotist finally gave up and reluctantly moved on to his next engagement, leaving the young man in his original predicament. Fortunately, a local hypnotist by the name of Al Layne had witnessed the event and offered to help. Reasoning that if Cayce could speak while under hypnosis, perhaps he could describe what was wrong with him, Layne easily put Cayce under and listened as he described the cause of his problem (a restriction in the blood supply to the vocal cords). When Layne suggested that Cayce attempt to increase the blood flow to the affected area himself, Cayce's face supposedly became flushed while his chest turned bright red. Upon awakening, Cayce's voice was said to have returned to normal—or so the story goes.

If he could diagnose his own illness while in a trance, Cayce wondered if he might be able to do the same for other people. Soon he began doing readings for others—usually free of charge—in which he consistently provided remarkably accurate diagnoses of his subjects' physical ailments. Armed with this new ability, which Cayce presumed to be a gift from God, he went on to become one of the most successful "diagnostic prophets" of modern times, successfully identifying and correctly prescribing treatment for literally thousands of patients and giving no fewer than 14,000 readings up to the time of his death in 1945—making him if not always the most accurate prophet in history, then easily the most prolific one.

Cayce was not without his critics. While he did demonstrate a remarkable ability to successfully diagnose illnesses while in a trance state (some claim an astonishing 90 percent success rate), he was also occasionally wrong, a point frequently noted by his detractors. It has also been charged that many of his diagnoses were so vague as to be little better than guesses (a charge commonly leveled against modern psychics as well). However, while it is a valid objection that needs to be considered, "guessing" does not explain how he managed to be uncannily correct in so many cases. Considering that Cayce had no formal medical training (though he was to eventually become extremely well-read on the subject), he should have been wrong more times than not. The fact that his success rate is considerably better than what one would expect by chance alone makes the notion that he was simply good at guessing simplistic.

Cayce's Prophetic Abilities

It was not just Cayce's knack for successfully diagnosing and treating illnesses that he was famous for, however. Beyond that rather remarkable ability, he was also known for his ability to supposedly foresee the future—especially regarding short-term events—reputedly with incredible precision. In fact, as the years went by, he became far more renowned for his prognostications—along with his teachings on Atlantis and reincarnation (two other manifestations of his many talents while asleep)—than for his skills as a medical diagnostician.

But how good a soothsayer did he prove to be? The results are mixed at best. Certainly, he did have some uncanny "hits," such as correctly predicting the stock-market crash of 1929 and the Second World War, but he also had many "misses" as well. We'll look at those in a moment, but first, the successes.

Probably one of his earliest and most poignant prophecies came in 1925, when the forty-eight-year-old Cayce, while speaking in a trance to a young doctor who had inherited a great deal of money and was wondering how to invest it, advised the man to exercise extreme caution and discretion in caring for his wealth in the face of "adverse forces that will come in 1929." On another occasion, in March 1929, a New York stockbroker was given a similar warning about an impending "great disturbance in financial circles" that was to take place soon. Six months later the stock market crashed, ruining thousands of investors and sparking the Great Depression of the 1930s, precisely as Cayce predicted.

His anticipation of the Second World War was equally impressive. Again speaking from a trance in 1935, Cayce warned that catastrophic events were building within the international community and went on to describe the entire world at war. He even managed to identify the key players, at least on the axis side, when he declared:

> *This will make for the taking of sides, as it were, by various groups or countries or governments. This will be indicated by the Austrians, Germans, and later the Japanese joining in their influence; unseen, and gradually growing to those affairs where there must become, as it were, almost a direct opposition to that which has been the theme of the Nazis (the Aryan). For these will gradually make for a growing of animosities. And unless there is interference from what may be called by many the supernatural forces and influences, that are activated in the affairs of nations and peoples, the whole world—as it were—will be set on fire by the militaristic groups and those that are for power and expansion in such associations...* [7]

Of course, the skeptic can always maintain that since Cayce's predictions came so close to the time of the events he described, he may have simply been particularly adept at "reading the times" and making especially good guesses. For example, the prospect of a downturn in the financial markets had been discussed throughout the 1920s as the superheated stock market reached unsustainable levels, making a crash almost inevitable.[8] As such, Cayce may have been unconsciously reiterating

---

7. Taken from http://www.edgarcayce.org/historychannel/cayce7prophecies.asp, a page on the Edgar Cayce's A. R. E. (Association for Research and Enlightenment) website (accessed June 11, 2009).

8. Cayce wasn't the only one good at foreseeing an inevitable crash on Wall Street. None other than financier Joseph Kennedy—ambassador to the United Kingdom and father of President John F. Kennedy—anticipated such a prospect and sold most of his stock just weeks before the crash, saving the bulk of his fortune and avoiding the financial ruin that affected so many of his contemporaries.

some of the speculation evident in his day (speculation he may have gleaned from his prodigious reading), making the fulfillment of his "prophecy" less remarkable than one might imagine.

Much the same might be made of his prediction about the Second World War. Certainly the prophecy would have been considerably more impressive had he made it a decade earlier, when the prospect of another world war coming on the heels of the great war of 1914–18 seemed remote. By 1935, however, war clouds were again building in Europe—especially in the aftermath of the ascension of the Nazis to power in Germany in 1933—making his prophecy less supernatural and more akin to an astute observation of the times. His inclusion of Japan in an alliance with Germany, however, would seem to challenge the odds of chance, especially considering that the two countries were not to formally ally themselves until 1941 and that in 1935 the prospect of an Asiatic war was far less plausible than a European war.

Predicting the stock-market crash and the Second World War weren't the only two correct prophecies Cayce made. He is also credited with foreseeing the end of global Communism, and that Russia would be reborn again (events that occurred in 1989–91). Interestingly, he also saw a strong religious movement coming out of Russia, though this is not yet evident. He even supposedly made a forecast in 1926 that anticipated the modern climatic phenomenon we know today as La Niña and El Niño. Speaking to a question about the future of wheat crops and weather patterns, Cayce made a connection between temperature changes in deep ocean currents and weather patterns when he said, "As the heat or cold in the various parts of the earth is radiated off, and correlated with reflection in the earth's atmosphere, this in its action changes the currents or streams in the ocean...."[9] Prophecy, a lucky guess, or did his prolific reading habits include Earth sciences and climatology? Difficult to know for sure but certainly interesting.

---

9. This quote appears online at http://www.edgarcayce.org/historychannel/cayce7prophecies.asp (accessed June 11, 2009).

Cayce's Earth-Change Prophecies

While one can argue whether Cayce really predicted the economic collapse of 1929 or truly foresaw the beginning and end of World War II, it's his end-times prophecies that proved to be among his greatest failures (a problem that likely resulted from Cayce's unfortunate habit of setting dates for his predictions rather than keeping them more general like most prognosticators).

Cayce's doomsday predictions generally fall into two broad categories: scientific and religious—the former having to do with massive Earth changes and the latter with Bible-based prophecies such as the Second Coming and the Battle of Armageddon (both of which Cayce predicted would occur in 1999). We'll look at his Bible-based prophecies in a moment, but for now let's look at Cayce's record in terms of how accurately he foresaw the planet's geological and seismic health.

Among the many prophecies Cayce made in terms of Earth changes were his predictions that between 1958 and 1998 there would be a series of dramatic earth changes that would result in " . . . inundations of many coastal regions caused by a drop in the landmass of about 30 feet combined with a melting of both polar ice caps," as well as " . . . the loss of much of England and Japan, the flooding of northern Europe, which will happen very rapidly." In other prophecies, he foretold a shift in Earth's magnetic pole in 2000 and of the destruction of Los Angeles, New York, and San Francisco, as well as the emergence of a new land mass " . . . appearing off the east coast of North America . . ." in 1968—all of which, of course, have proved to be erroneous.

His most ardent supporters, however, maintain that at least some of these "misses" may not be misses at all. For example, it is claimed that his prediction that Earth's poles would begin to shift in the year 2000 is potentially being realized, albeit slowly. In a 2003 episode of the PBS series *Nova*, "Magnetic Storm,"[10] scientists maintain that the

---

10. Written and produced by David Sington. See http://www.pbs.org/wgbh/nova/magnetic/about.html for a program description and transcript (accessed June 11, 2009).

vital magnetic shield that protects the planet from all types of cosmic radiation may be weakening in anticipation of a pole reversal (something the planet does about every 250,000 years or so). The *Nova* program also revealed that the shift has indeed begun in the South Atlantic Ocean region, where the north-south polarity is fluctuating back and forth, thereby weakening the magnetic shield against solar radiation. Apparently this was first noticed around the turn of the millennium, roughly when Cayce said it would begin. Only the fact that it may take centuries for the effects to become obvious and deleterious to humans prevents it from being as well known as it might otherwise be, which is why the prophecy is often overlooked, at least according to Cayce's proponents.

Another prophecy that may or may not have been successfully realized was his 1936 prediction that the mountaintops of the lost continent of Atlantis would reappear in the Bahamas near the island of Bimini " . . . sometime in 1968 or 1969." Sure enough, in 1969 divers located an unusual underwater formation of limestone boulders in the shallow waters off the island that resembled a massive *J*-shaped road nearly half a mile long. Initially touted as evidence for the lost continent due to the eerily symmetrical shapes of the stones (implying that the structure was artificially created when the shallow sea floor around the Bahamas was last above sea level nearly 12,000 years ago), scientists later decided it was a natural formation no more than four thousand years old. While debate continues to this day (some of the stones do appear to be almost perfectly square and to have been carefully stacked upon each other), it will probably never be possible to determine with any certainty whether the formation in the waters off Bimini are just a bunch of rocks or evidence that, once more, Edgar Cayce may have been far more astute a prophet than some are willing to concede.

Cayce's Bible-based Doomsday Prophecies

While Cayce made a number of often dire Earth-change predictions that went either unrealized or are open to debate, it was his Bible-based doomsday predictions that are most disappointing (or fortuitous, considering that they didn't occur as predicted). Possibly foreseeing the various catastrophic Earth changes just mentioned as a harbinger of the end, he unwisely implied that the epic Battle of Armageddon and the Second Coming of Christ would occur in 1999—an implication that, despite being made over half a century ago, still manages to sound very contemporary.

What are we to make of these misses of Cayce's? Are they evidence of a fantasy-prone personality or was Cayce cleverly making things up as he went along? Most likely, the latter possibility can be discounted, as there is absolutely no evidence to suggest that Cayce was a con man, as evidenced by the fact that he rarely charged for his readings and lived out most of his life in comparative poverty when he could have easily used his abilities to enrich himself many times over. And as for possessing a fantasy-prone personality as many of his critics maintain, that too appears problematic. Cayce appeared to be very down-to-earth and astute to those who knew him personally, and even, on occasion, quite skeptical of his own utterances. He may have been a bit "eccentric" by modern standards, but there is nothing in his writings that imply the man was in any way delusional.

So what's left? If he was not a kook or a con man, how could he have been so extraordinarily wrong about the end-times and Earth changes when so many other predictions of his proved to be spot on?

This may be where Cayce's deep-seated religious faith could have hindered him. What many people who have heard of Cayce don't know about the man is that, despite his teachings on what are usually considered unorthodox or even New Age concepts, at his core he remained profoundly Christian throughout his life. (Cayce is said to have read the entire Bible cover to cover every year of his adult life, and he even taught Sunday school classes.) In fact, the apparent contradictions between his trance-induced statements and his orthodox Christian beliefs often proved to be a source of considerable consternation to him—particularly with regard to his belief in reincarnation and its conflict with the traditional church's teachings on a single lifetime. In spite of all this, however, Cayce remained a professed Christian all his life and consistently managed to integrate his beliefs into his prophetic utterings.

But could his deep-seated faith have been the reason for his spectacularly failed prophecies concerning the Second Coming? In effect, could Cayce have been unconsciously victimized by his own background as a student of biblical prophecy? The possibility that Cayce was so heavily influenced by the Christian-based doomsday belief systems of his day that he allowed them to cloud his judgment, resulting in him inadvertently trying to use his prophetic gifts to confirm his own preconceived opinions by massaging his prophecies to conform to his deep-seated religious beliefs, has to be considered. Certainly, he wouldn't be the first person to be blindsided by his own belief system, nor will he be the last.

While it's difficult to know with any degree of certainty, it does appear that Cayce's prophecies proved to be less accurate the more they conformed to orthodox Christian constructs. Is it possible, then, that Cayce may have been a man who possessed something of a religion-fueled imagination *in addition to possessing a genuine gift for prophecy*, and that the two sometimes became confused, resulting in him making some truly preposterous end-times predictions? This dichotomy is possibly what gave him both his impressive list of successes as well as his most glaring "misses"—making him, if not a "sleeping" prophet, certainly an inconsistent one.

Conclusions

That Edgar Cayce was a man who had some remarkable abilities cannot be denied. However, we also can't ignore the fact that his predictions just as frequently missed their mark by miles, which leads us back to our original query as to whether Cayce truly had a gift for prophecy.

Unfortunately, as is the case with most prophets, "hits"—that is, predictions that appear to have been realized historically—are more readily trumpeted than are the "misses," which tend to be quickly forgotten, explained away, or otherwise denied, and Cayce is no exception. But such is the norm with the prophecy game and has been from the beginning, which is what makes it so difficult to determine whether there is a genuine phenomenon afoot or if some people are just especially astute observers. In spite of all this, however, Cayce remains an enigma to this day and one who continues to fascinate students of prophecy. Certainly, irrespective of his abilities to frequently successfully diagnose various ailments while in a trance, his influence on modern Western beliefs with regard to things like reincarnation and Atlantis has been profound, and will likely remain influential for years to come—at least, until the next "sleeping prophet" arises to take his place.

Chapter Six

2012 Hysteria

Having briefly looked at the less-than-illustrious history of failed prophecy in the past as well as having examined two of prophecy's biggest "stars," we are finally ready to examine modern prophecies in some detail. Becoming aware of how spotty the record has been to date and how subjective and often emotion-driven the entire issue can be will hopefully make us a little more careful about embracing doomsday predictions of the future too quickly. As in all things, caution must be the keyword here.

No date in recent history has become as popular with end-times proponents as has the year 2012, which we touched upon briefly in chapter 1. Why this particular year has proved to be such a lightning rod for end-times proponents is an interesting story in its own right, especially since it did not come into our modern consciousness until comparatively recently, making it something of a Johnny-come-lately in terms of prophetic date-setting. However, despite its fairly recent introduction to the apocalyptic pool, it has grown in popularity beyond anything its Y2K predecessor enjoyed, and appears to be gaining popularity—and adherents—with astonishing speed. But where did it come from, and why do so many people see this particular year as so significant?

Blame it all on the Maya.

The Maya and Their Amazing Calendars

It all seems to have started with the belief that, according to Mayan teaching, the world will end on December 21st (or 23rd, depending on your source) of 2012—a date arrived at based upon the fact that the Mayan calendar ends its fifth 5,125-year cycle on that date. Such an abrupt ending, then, is perceived by some to mean that the winter solstice of 2012 is the moment when the Maya believed the world will end—or, more correctly, go through a period of cataclysmic change. What sorts of "change" this will entail is anybody's guess, of course, but it is taken by many to be a harbinger of disaster. Not surprisingly, then, that makes the date a major source of worry for millions of people around the world.

The fact that the Mayan "long count" calendar ends on that date has been known to archeologists for decades, but it never seemed to generate much interest until it was written about by an art history and aesthetics professor from the University of Chicago named José Argüelles. In his book *The Mayan Factor: Path Beyond Technology* (first published in 1987), Argüelles writes that the ancient Maya, a pre-Columbian Native American culture that flourished for nearly two thousand years in Central America, had figured out through precise astronomical calculations that the earth (or, technically, the earth's "fifth sun") would end at the winter solstice, December 21, 2012, at which point a new, sixth 5,125-year cycle would begin.

Now Argüelles, along with others, decided that when this occurred, all the evils of the modern world—war, materialism, violence, injustice, governmental abuse of power, and so forth—would end with the proverbial bang, at which point the sixth sun and the fifth earth would dawn to a new age of world peace. So convinced of this belief was Argüelles, in fact, that in August of 1987 he initiated something called the "Harmonic Convergence"—an event that saw people from all over the world (known as "lightbearers") gather at various sacred sites and "mystical" places on the planet—in an effort to usher in a new era of peace and officially start the final twenty-five-year countdown to the end of the Mayan Long Count calendar in 2012.

While the book he wrote garnered some interest within the New Age community, it was this "convergence" event that received the lion's share of media attention and was instrumental in kicking off the entire controversy that ultimately made 2012 the official "end of history" date within popular culture. As a result, the Mayan prophecy has become so popular so quickly that where there was practically nothing written on the subject just ten years ago, as of this writing there are dozens of titles on the bookshelves having to do with the Maya and their bizarre calendar—each either reinforcing Argüelles' original concept or putting its own unique spin upon it. No doubt about it, the Maya and their mysterious calendar is *the* hottest thing in prophecy right now, and will likely remain so right up until Christmas Day, 2012.

So what is it about the Mayan calendar and the magical date of 2012 that seems to have so many people excited? To answer that question, let's take a moment to examine the Maya and their calendars to see exactly what they were talking about. After all, a doomsday prophecy only has to be right *once*, making it worth taking a closer look at the Maya and their mysterious calendar, if only for our own safety.

Who Were the Maya?

Most people have at least heard of the Maya, even if they don't know much more about them than that they lived in Central America and had this obsession about calendars. But who were they exactly?

The Maya are an indigenous people who populated the area we know today as southern Mexico,[1] Guatemala, and El Salvador from roughly 1800 BCE until around 900 CE. Although the Maya flourished in the Americas for over two thousand years, their "golden age" occurred between 250 and 900 CE, at which point they were one of

---

1. It's ironic to me that Mayan civilization flourished throughout the Yucatán Peninsula—the very site where the K-T asteroid struck sixty-five million years ago, wiping out most life on the planet, including the dinosaurs. More than a coincidence? Probably not.

the most densely populated and culturally dynamic societies in the world. Known for possessing the only fully developed written language in pre-Columbian America, the Maya were also famous for their spectacular art, monumental architecture, and sophisticated mathematical and astronomical systems—making them, in many ways, as advanced a society as any that existed in Europe at the time. They even built irrigation systems and developed building techniques as dynamic and sophisticated as those of ancient Rome that could even be considered almost modern by our standards today. Then, for reasons still debated among archeologists, around 900 CE they abruptly abandoned their great cities and, though they never entirely disappeared (the Mayan language is still spoken in some parts of Mexico to this day), by the time the warlike Aztecs came along in the late 1200s, they had all but gone, leaving only crumbling but still magnificent ruins to mark their presence.

It was their advanced understanding of mathematics and astronomy, however, that is the most impressive aspect of these remarkable people. The Maya were famous for producing a series of calendars that track, with remarkable precision and accuracy, various lunar, solar, and earth cycles. These calendars act like a type of harmonic calibrator, linking and coordinating the earthly, lunar, solar, and galactic seasons in an aesthetically simple and elegant manner. In fact, their calculations proved to be so extraordinarily accurate that they are capable of impressing us with their skills even today, and it is these calendars that have end-times aficionados intrigued and remain the source of so much speculation.

How they work is fairly complex, but not indecipherable. Basically, the Maya used three different dating systems in parallel: the *Long Count*, the *Tzolkin* (divine calendar), and the *Haab* (civil calendar). I won't go into all the details about how each works, but suffice it to say that they each use very different dating cycles from our modern Gregorian calendar. For example, the Tzolkin has weeks of two different lengths, one a numbered week of thirteen days and the second a week of twenty named days, while the Haab calendar consists of eighteen months of twenty days each, followed by five extra days, known as

Uayeb, giving the year a length of 365 days. Long time periods are measured by means of the Long Count, in which one 360-day year (a *Tun*) consists of eighteen twenty-day months (known as *Uinals*). Twenty of these Tuns makes a *Katun*, twenty Katuns a *Baktun* (nearly 400 years), and thirteen Baktuns a "Great Cycle" of 1,872,000 days, or about 5,125 years. Got it? Me neither.

The problem comes from the fact that, at least according to the Maya, there are five Great Cycles in all, which when combined denote the history of time (from their perspective). Scholars who have studied this calendar in detail have been generally able to agree that, as best they can determine, the last Great Cycle began on the 11th of August in 3114 BCE, and is set to end on the 21st of December in 2012 CE (13.0.0.0.0 in the Long Count), thereby finishing off the Maya's 26,000 year, five-cycle calendar. Now, what's most interesting about this is that at around 11:00 PM Greenwich Mean Time on that date there will be an extremely close conjunction of the winter-solstice sun with the crossing point of the Galactic Equator (Equator of the Milky Way) and the ecliptic path of the sun. This is something that happens once every 26,000 years, making it a big deal.

Argüelles' implication is that the end of this cycle was interpreted by the Maya as being synonymous with the end of time, or at the very least a period marked by major changes to the planet, both physical and spiritual. However, there is no evidence that the Maya believed that at all. While doubtlessly the Maya would have interpreted this to be a significant event, that doesn't necessarily mean they considered it a time of catastrophic changes. They would have seen spiritual significance to be sure, but there is nothing to suggest that the Maya considered such a change to be apocalyptic in nature.

To be fair, Argüelles did not teach a doomsday scenario, but interpreted this change of cycles as a time of greater understanding that would be instrumental in ushering in a period of world peace. Like the Harmonic Convergence of 1987, it would be unlikely to result in any dramatic shifts in human enlightenment overnight, but would simply mark the beginning of a rapid spiritual advancement among all human beings. It was for later writers to suggest a more traditional doomsday scenario, thereby inadvertently turning what was intended as a positive interpretation of 2012 into the darker and more frightening connotation it holds today.

Unfortunately, it appears to be a common human trait to interpret such specific date-setting in generally apocalyptic terms, with the result that many people see the end of the Mayan calendar and December 21, 2012, as doomsday rather than just the last day on an ancient calendar. What effect this tendency is likely to have over the next few years as the date approaches remains to be seen, but it could well be significant. Will it bring people together in a sense of anticipation for a brighter future or result in panic on a global scale? If the past is any indication, we should expect to see evidence of both to be followed by a collective sigh of relief as the date passes without incident. Only time will tell.

On the other hand, if the Maya are on to something it would imply that the cosmos are a very orderly place in which even something like the apocalypse can be predicted with precise accuracy. Of course, in the case of a total apocalypse that wouldn't do humanity any good, but it would be nice to know that God runs a very orderly universe.

Other 2012 Scares

What Argüelles started with his book in the 1980s has had an unforeseen and probably unintended effect, and that is to make 2012 synonymous with doomsday. In effect, once the Mayan date became part of our collective consciousness, the year 2012 began serving as a catalyst for other doomsday proponents.

This result, however, is neither without precedence nor entirely unexpected. After all, what better way to attract attention to one's own prophetic musings than to tie them to an already popular end-times date? This is why the last few years have seen a burgeoning of doomsday prophecies, all predicting the end (however such is defined) in or as close to 2012 as possible, and why the popularity of the Mayan date is likely to continue to serve as a type of lightning rod around which other doomsday predictions will continue to conglomerate.

The difference between many of them as compared to the Mayan predictions, however, has to do with the specifics of doomsday. Whereas proponents of the Mayan prophecy tend to be vague about precisely *how* the planet will be affected in 2012, others are considerably more precise, declaring that the end will come as a result of anything from a sudden polar shift to a dramatic and lethal increase in sunspot activity (which is supposed to result in all sorts of disasters, from massive tsunamis to increased volcanic activity to bursts of especially intense cosmic radiation). This tendency by various prophets of doom to find ways to get their own particular scenarios to fit the date—or, at least, the general time frame—has had the effect of giving the general public the impression that numerous unrelated prophecies are all converging at a particular point, thereby suggesting some sort of vast cosmic doomsday collusion is afoot.

Part of the problem, however, is that there are elements of truth to at least some of the predictions. For example, it is a fact that 2012 is slated to coincide with a period of especially intense solar sunspot activity. Some scientists have even suggested it may be the most intense cycle of sunspot activity in fifty years, though this remains somewhat speculative. Sunspot activity always increases every eleven years as part of the sun's natural cycle, after which it will die down for another eleven years before starting the cycle over, so this is nothing new. What is new is that now this cycle is seen as something ominous rather than natural, with all the doomsday imagery that entails.

The facts are somewhat less spectacular than some would have us believe. While increased sunspot activity is known to interfere with radio communication and occasionally wreak havoc on orbiting satellites and spacecraft, it has little impact on the planet itself. The only conceivable impact an especially robust sunspot cycle might have on Earth is this: since the sun is slightly hotter during high periods of sunspot activity (though by only a tiny amount, on the order of 0.1 percent of the solar constant), there is some concern this will further aggravate global warming. This is really only a minor consideration, however, as it is the level of greenhouse gases in the atmosphere, not sunspots, that appears to be the more important element in determining overall planetary warming.

Something else that occurs every eleven years that has doomsday proponents excited is that the sun's magnetic pole flips from one pole to the other. This last occurred in 2001 and, according to NASA scientists, is slated to do so again sometime in 2012. As such, some doomsday aficionados have claimed that when this happens, it will place such a pressure on our own planet's magnetic poles that it will cause them to likewise shift—that is, compasses will no longer point toward magnetic north but will point south—causing massive geothermal and tectonic catastrophes on the planet. Some have even gone so far as to suggest that this shift will induce Earth to likewise shift poles entirely—effectively resulting in the planet rolling over onto its back and reversing its orbit—which really would have most unpleasant consequences to all living things.

Neither theory, however, is based in scientific fact. First, while the earth's magnetic pole does occasionally shift every once in a great while (the last reversal occurring about 740,000 years ago), there is no evidence that such a shift is imminent, nor is the impact upon the planet such a shift would incur particularly well understood. Clearly, such a shift in the poles would cause problems on Earth, but how catastrophic it might prove to be to human life is open to debate.

As for the entire globe flipping over and/or reversing its orbit, however—which really would be catastrophic—such a prospect is, from a purely scientific basis, impossible. The earth is a giant gyroscope operating in a weightless vacuum; to get it to stop spinning on its axis (or even just to slow it down a bit, much less reverse its orbit) would require an enormous amount of energy. The only thing even remotely capable of performing such a feat would be a glancing blow from a moon-sized planetoid that got close enough and was at the precise angle to overpower the planet's current momentum to get it to reverse its spin. Of course, such a near-miss would strip the planet of its atmosphere and probably sheer off trillions of tons of material as it passed by, making the reversal a moot point as it would be unlikely that there would be anyone left alive on the planet to see it.

Another potential planet killer waiting in the wings is something called—unimaginatively enough—Planet X (not to be confused with the 1950s sci-fi movie planet of the same name). While *Planet X* is scientifically valid nomenclature for any yet-to-be-discovered, potential celestial object implied by aberrations in the orbital paths of outlying planets (such aberrations in Neptune's orbit were the impetus behind the discovery of Pluto in 1930), the Planet X of doomsday lore is more sinister. It is the brainchild of an Azerbaijani author by the name of Zecharia Sitchin—a largely self-taught and self-professed expert in modern and ancient Hebrew, various Semitic and European languages, the Old Testament, and the history and archeology of the Near East—who theorized in 1976 about the existence of a twelfth planet (which he maintained the ancient Sumerians referred to as *Nibiru*) lying just beyond the orbit of Pluto. While that was an extraordinary claim in its own right, Sitchin further contended that this remarkably Earth-like planet[2] possesses a massively elliptical orbit that allows it pass close to Earth every 3,600 years. During these close flybys, the highly advanced residents of this planet would interact with the natives of Earth in various ways.

According to Sitchin, these entities first arrived on Earth some 450,000 years ago and created humans by genetically engineering female apes, and they have been returning periodically ever since to continue tweaking their creation. Apparently, Nibiru is due for another visit anytime now (perhaps in 2012?), when it is imagined that, largely due to the more advanced level of technological sophistication we possess today (as opposed to that of our ancient predecessors on the last visit), all hell will break loose. Since the Nibiruans are supposedly still thousands of years more advanced than we are, however, such a clash would probably look more like doomsday than a gentle exchange of genetic materials.

---

2. Nibiru is supposed to be able to remain Earth-like when far beyond the orbit of Pluto due to being heated from within by radioactive decay.

Of course, the problems with Sitchin's theory are many. Apart from his shaky ability to accurately translate and correctly interpret Sumerian cuneiform is the complete lack of any evidence to support his claims and the physical impossibility of a planet with such a massively elliptical orbit being capable of sustaining life when it spends 99 percent of its time in darkness beyond the orbit of Pluto. Additionally, if his hypothesis is correct, it seems we should see some evidence of Nibiru's last visit around 3,600 years ago in the archeological record, but beyond trying to suggest that references to the Nephilim and Seraphim in the Old Testament were mythicized accounts of our Nibiruan creators, such evidence remains conspicuously absent. Still, Sitchin has his proponents, who continue to maintain his twelfth-planet scenario with a tenacity that borders on the fanatical.

Others have taken Sitchin's idea about a still-missing planet down a somewhat more plausible path by proposing the existence of a hypothetical planet—one at least as large as Jupiter—existing just outside our known solar system, not far from the Oort cloud. Such a planet would not be inhabited, of course, and it wouldn't really pose much of a threat to us unless it were large enough to disturb the orbit of long-period comets, thereby potentially spitting one our way every once in a while. Others have suggested that this mysterious planet may not even be a planet at all, but something even larger and more deadly: a brown dwarf sun, which, if true, would not only make our sun part of a binary system but might affect our own sun in mostly detrimental ways.

Some have suggested that were such a star to exist, it might have the capacity to increase the temperature of our sun, thereby causing an increase in the severity of the next peak in the eleven-year sunspot cycle in 2011–12. This, in turn—assuming it were severe enough—would have the effect of potentially destroying the earth's magnetic field, which really would be unfortunate. Not remarkably, the prospect that there is a mysterious unknown planet/brown dwarf star out there threatening our planet is not enthusiastically championed by orthodox science, largely due to the fact that to date the hypothesis lacks any evidence to support it.

The Deadly Comet of 2012 Foretold by the Bible Code

But perhaps the most interesting cataclysmic prediction comes from the Bible itself, though not in the form of a prophetic utterance contained within the apocalyptic writings of the Book of Revelation as is normally the case. This one is the result not of *what* the Bible says but in *how* the words of the Bible are arranged. According to journalist Michael Drosnin, author of *The Bible Code*, the Bible should not be approached as an oracle of ancient prophecy but as a type of cryptograph containing hidden messages within the jumble of words in some of the Old Testament texts.

By using something called Equidistant Letter Sequencing (ELS), Drosnin contends that one can find meaningful and related patterns of words and dates in close proximity to each other within the words of the Pentateuch (the first five books of the Old Testament and the heart of the Jewish Torah) that would seem to go beyond mere chance. For example, by taking every fifth letter in a diagonal line of text to create an understandable word or phrase and then affixing it with prophetic meaning, he has apparently found all sorts of references to major historical events such as the JFK assassination and 9/11. Claiming the odds against such a combination of words appearing in the randomness of the Pentateuch's tens of thousands of letters in such close proximity to each other to be "astronomically high," Drosnin is convinced he has found a true prophetic code that can be uncovered purely by using state-of-the-art computer software to search for corresponding letters and relevant words.

And it's not just past events that Drosnin has uncovered, but hints at future catastrophes as well. To that effect, he has gone on record as claiming to have found a hidden message in the Pentateuch that predicts that a comet will crash into the earth in 2012 and annihilate all life, forcing him to the conclusion that 2012—at least according to his process—*really is* the year of doomsday.

However, critics have dismissed Drosnin's methodology as little more than a parlor trick, demonstrating that meaningful words and phrases can be produced using his method on any similarly large manuscript.

For example, Australian mathematician Brendan McKay, an ardent critic of Drosnin's process, demonstrated that a computer search of Herman Melville's nineteenth-century classic *Moby Dick* found a number of meaningful phrases in close proximity to each other. Perhaps most significant was McKay's discovery in *Moby Dick* of a series of words relating to Israeli Prime Minister Yitzhak Rabin's assassination in 1995 that contained both the assassin's first and last names, the university he attended, and even the motive for the attack ("Oslo," relating to the Oslo accords of 1993, which created the Palestinian National Authority and granted it partial control over parts of the Gaza Strip and West Bank).[3]

Although later rebutted by Drosnin and his supporters, McKay did appear to demonstrate successfully that just as the eyes can be tricked into seeing familiar faces in random patterns of light and shadow, so too can the mind be tricked into finding meaningful phrases where none exist in random collections of letters. Drosnin's critics also point out that Drosnin, in his 2002 sequel, *Bible Code II: The Countdown*, described the probabilities of a nuclear holocaust and the destruction of major cities by earthquakes in 2006, claiming that "the danger will peak in the Hebrew year 5766 [September 2005 to September 2006 in the modern calendar], the year that is most clearly encoded with both 'world war' and 'atomic holocaust' . . . "[4] That prediction, of course, turned out to be erroneous[5] and makes his 2012 comet-strike prediction equally suspect.

This doesn't necessarily demonstrate that a comet *won't* strike the planet in 2012, of course, but it does force us to pause when considering such dire predictions and wonder how responsible it is to frighten

---

3. That example and many more can be found online at http://cs.anu.edu.au/~bdm/dilugim/moby.html (accessed June 12, 2009).

4. Michael Drosnin, *Bible Code II: The Countdown*. (New York: Penguin, 2003), 20.

5. In fairness, Drosnin does say that " . . . the Bible code is not a prediction that we will all die in 2006. It is a warning that we *might* all die in 2006, if we do not change our future" (*Bible Code II*, p. 237), making his predictions hypothetical rather than predictive in nature.

people unnecessarily. Certainly, if the Bible genuinely does suggest annihilation is just around the corner, we must wonder why such an event would be so thoroughly hidden within the words of an ancient manuscript. Wouldn't the supposed author of the Pentateuch—assumed to be God Himself (or, as proposed by others, extraterrestrials from the distant past)—be a bit more clear about it? After all, it *is* the extinguishing event of human history we're talking about here, making it, one might argue, something we would be well served to know about. Would God (or the ETs for that matter) really be so cavalier about such a thing?

Conclusions

I have barely scratched the surface of the many yet-to-be-realized (or not realized) prophetic beliefs currently in vogue, but this small sampling should be sufficient to demonstrate that the next few years promise to be every bit as eventful as far as prophecy goes as any time in the past, with every war, category-five hurricane, and political upheaval being interpreted by someone somewhere as the beginning of the end.

Of course, only time will tell if any of them pan out or if 2012 really will be a pivotal year in the history of humanity (or, for that matter, the *final* year of human history). Obviously, if you are reading these words after that date with no hint of the apocalypse present, that should tell you something. Time, it turns out, has brutalized many a prognosticator by managing to ignore even the most carefully researched doomsday date. I suspect that time will do the same again, but with even greater vengeance as the magical 2012 date comes and goes with the normal allotment of wars, natural disasters, and crises being evident. Whether the prophets of doom will finally surrender and admit that their predictions are nothing more than wild and sometimes irresponsible guesses remains to be seen, but I suspect that they will simply reinterpret their charts, reread the signs, or receive further visions that will point them toward new dates further down the road. This seems to have been the pattern in the past and is likely to remain so in the future, for as long as there are people bold enough to keep making such predictions and no shortage of people willing to believe them, doomsday will always remain just a day away.

Chapter Seven

Prophecy in the Bible

I was just sixteen when I was first introduced to Bible prophecy.

It all started in the summer of 1974. My brother-in-law, Charles—a man who had unwittingly become something of a mentor and spiritual advisor to me during the darkest days of my interminable adolescence—introduced me to a book entitled *The Late, Great Planet Earth*, probably never for a moment realizing what a profound impact the dog-eared little paperback would have on me. Written by an up-to-then little known tugboat captain turned prophecy expert named Hal Lindsey, *Late, Great Planet Earth*[1] (henceforth known as *LGPE*), was what one might call a "doomsday" book in that it dealt almost exclusively with the end of the world (or, more precisely, the end of the present age).

Such books had appeared before, but none went into the kind of detail Lindsey went into as he described in a highly readable and even entertaining way how Russia and her allies would soon be sweeping down into the Middle East to exterminate Israel, only to be destroyed by a federation of European states (*the United States of Europe*, he

---

1. The book was to become one of the best-selling nonfiction titles of the 1970s, eventually being translated into fifty-four languages and selling over thirty-five million copies. In fact, it proved to be so popular that it remains in print to this day, an astonishing thirty-nine years after it first appeared. Not bad for a first-time author.

called it in the pre-European Union days of 1970) under the direction of some super leader known as the Antichrist. He then went on to describe how after successfully dispatching this cabal of Communist nations and their Arab allies, a short period of comparative calm would ensue before a new enemy—this time much larger and more formidable than the Soviet-led armies proved to be—would arise "in the east" (Lindsey decided these were the Communist Chinese) to challenge the Antichrist's armies to a final showdown that would make World War II look like a church social.

His narrative of this final engagement between a quarter of a billion men locked in mortal combat was as gory as it was spectacular. (Lindsey describes blood flowing as high as the bridles on the horses; I'm not sure why there are horses on this futuristic battlefield, but Lindsey seemed pretty sure there would be.) To a kid who already possessed an unhealthy obsession with military history to begin with, this was irresistible.

But that wasn't the whole story. Lindsey also described how, prior to the great Soviet/Arab invasion of Israel, all the Christians in the world—or, at least, all the "real" Christians[2]—were going to suddenly and instantaneously disappear from the face of the earth via something known as the "rapture," an event that will usher in a seven-year period of terrible tribulation that will leave millions and, potentially even billions, of people dead (with many of the survivors *wishing* they were dead).

Fortunately, *LGPE* has a happy ending—at least for those who profess Christ as their savior and manage to survive the rampant persecution evident during this dark time: Jesus Christ, along with all those Christians originally snatched off the planet seven years earlier, return to destroy the Antichrist and his followers and reestablish his throne on Earth. No doubt about it—Hal Lindsey knew how to spin a tale.

Not remarkably, I bought into his scenario, not just because I thought it was an exciting prospect, but because, at least according to

---

2. "Real" here being a code word, it is thought, for those who believe as Lindsey does, though he does not explicitly state this in his writings.

Lindsey, all this stuff was in the Bible. Now even to a nominal Catholic with only the most perfunctory knowledge of Scripture, that was a big deal. I may have never read a complete chapter of the "good book," but if it was in there it *must* be true, demonstrating that my faith in things I knew practically nothing about was impressive, even by modern standards.

So did all this conspire to get me into the church? Well, not right away. I really wasn't "into church" as a teenager, even if joining held out the promise of sparing me from having to endure the horrific fate that was in store for so many of my fellow humans. Yet *LGPE* and the fantastic imagery it painted never left me. In fact, it continued to haunt me for the next few years until it finally became such a source of concern that, shortly after reaching my twenty-first birthday, I decided to get serious about all this "Christianity stuff," and became a full-fledged, card-carrying born-again Christian. Not only did I now consider myself to be "saved," but I was also instantly eligible to be among those who would be "raptured" when the time came, which I assumed to be—at least according to Lindsey's implied timeline—fast approaching.

The desire to avoid the tribulation period was not my sole reason for becoming a Christian, but it did play a significant role. At the time I saw it as one of the many tacks God was using to ensure my eternal salvation, and I was grateful that He thought enough of me to go through so much trouble to rescue me from my sins, both real and imagined. Encouraged by such generosity on His part, I finally began reading that Bible I'd heard so much about, along with about every commentary and faith-based book I could get my hands on, until I eventually became reasonably well-versed in the Bible—at least in comparison to most churchgoers.[3] I was no theologian, to be sure, but after a while I knew my Nehemiah from my Thessalonians with the best of them, and had even committed entire chapters of the New Testament to memory.

---

3. I'd heard somewhere that only about 5 percent of professing Christians have ever read the entire Bible through from cover to cover. I don't know if that figure is true, but in my experience if it's wrong, it's not off by much.

(This was back when I still had a memory that could do things like that.) In the process, I not only made myself increasingly familiar with Lindsey's subsequent works on the subject,[4] but I don't think it's too great a stretch to maintain that I became as much an authority on biblical prophecy in general as anyone without a day of formal seminary training could be.

There was one problem, however: while I was relieved to have made it to safety by the proverbial skin of my teeth, I was saddened by the fact that most of my friends and family wouldn't be joining me in my great ethereal adventure. Naturally, this propelled me to embark on an annoying and spectacularly unsuccessful effort at converting them to my theological perspective, which left me with a sense of frustration. It seemed that despite my best efforts, there was nothing I could do to save them from their fate. All I could do was hope they would be wise enough to convert once I and millions of other like-minded Christians vanished in the blink of an eye, thereby demonstrating the truth of God's Word and their need to repent. That was, after all, the whole point of the seven-year tribulation period. It was specifically *designed* to get people to repent, and even though it was lost upon me how God intended to demonstrate His great love for humanity by, in effect, torturing it into repentance, I had faith He knew what He was doing.

And perhaps that's where all the trouble started. As I grew older, I began to question the belief system I had so blindly embraced—not just because of my concerns for my family and philosophical questions over the rationale behind the tribulation period, but also because as time passed I began to notice that Hal Lindsey's timeline seemed to be askew. He claimed—as did every prophecy expert worth his or her salt—that Israel's reestablishment as a nation in 1948 was the pivotal event that effectively started the doomsday clock, so to speak. It's a little difficult to

---

4. *Satan Is Alive and Well on Planet Earth* (Grand Rapids, MI: Zodervan, 1972), *There's a New World Coming* (Santa Ana, CA: Vision House, 1973), and *The 1980s: Countdown to Armageddon* (New York: Bantam, 1980), among others.

explain, but it seems that once Israel became a sovereign nation after two thousand years of wandering, the final biblical "generation" supposedly spoken of by Jesus in the gospels (Matthew 24:34) was in place, implying that Jesus was to return within forty years (a "generation" as figured by most Bible scholars). By my reckoning—and implied by Lindsey if not precisely stated—that meant that Jesus would be returning sometime in 1988. If correct, and being that Jesus' return was to be preceded by a seven-year period of tribulation, that put the rapture squarely in the middle of 1981, which, considering that I had only become a Christian a couple of years earlier, made the rapture not only a theoretical possibility to me but also a very real and urgent event I anticipated experiencing any day.

Unfortunately—or fortunately, as the case may be—1981 came and went without a single Christian being raptured or any sign of an emerging Antichrist, which I found to be a little perplexing. Lindsey's scenario seemed so well thought out and precise that I couldn't understand what the hold-up might be. However, I did have confidence that the Bible—and, by extension, Bible scholars—knew what they were talking about, so I decided to just sit tight and wait it out.

Eventually I came to the conclusion that the 1981 rapture date might have been presumptuous. Perhaps Jesus was speaking more metaphorically when he used the term *generation* and was not suggesting that it would be *precisely* forty years. Additionally, some preachers suggested that Jesus was intentionally delaying his return, perhaps in a compassionate effort at giving every last person on the planet one final chance to embrace salvation (although one would think that for the prophecy to be particularly useful, the Savior would need to stick to the original timeline). And the Soviet Union did still pose a threat—as did Communist China—so I wasn't prepared to dismiss all of Lindsey's prognostications out of hand, but it did seem odd that such a precise timeline could have been so inaccurate. Nonetheless, I continued to believe that the rapture—and with it the hoofbeats of the coming apocalypse—couldn't be too far away.

By the end of the eighties, however, things really began getting confusing. First, despite it being the fortieth anniversary of Israel's establishment, 1988 also came and went without a peep. Then, to my even greater surprise, in 1989 the Berlin Wall came down and the Soviet Union, that great bogeyman of both the Cold War and Hal Lindsey's doomsday scenarios, quietly exited the world stage; the catalyst that was to kick off the entire course of events—the Soviet invasion of the Middle East—became increasingly unlikely, further bringing Lindsey's theory into question.

Of course, it wasn't all bad news for Mr. Lindsey. Communist China still had the potential to mobilize a two-hundred-million-man army capable of invading the entire Indian subcontinent, and progress was being made in terms of unifying Europe into a single political, economic, and military entity, thereby setting the stage for the emergence of the Antichrist. However, even I had to admit upon closer consideration that neither prospect appeared likely: the idea that China (or any country on the planet, for that matter) could sustain an army of two hundred million men in the field—much less move it seven thousand miles across the Himalayas, the vast Indian subcontinent, and the barren deserts of Iran and the Middle East—appeared increasingly untenable.

As for the much ballyhooed European Union—well, let's just say it sounded far less ominous than it had just a decade earlier. Lindsey said it would be a ten-nation confederacy headquartered in Rome and that it would be one of the most powerful world governments ever known, but instead it appears to be—at least as of this writing—little more than a bunch of bickering European governments sharing a common currency in an effort to compete economically with America and Asian nations. Furthermore, the modern European Union has considerably more than ten members, isn't headquartered in Rome, and I don't see that it speaks with a single voice or has anything approaching an *Überführer* at the helm ready to spark WWIII. If there is an Antichrist in the mix, he is extraordinary inconspicuous.

And so I suffered through a growing suspicion that something was not right with my eagerly embraced belief system. Expanding my reading list to include Christian scholars that viewed biblical prophecy as being largely allegorical in nature or already historically realized in the distant past—positions I had previously considered heretical—pulled me even further from Hal Lindsey's position until I eventually came to the conclusion that he and others like him didn't know what they were talking about.

Unfortunately, this had a cascading effect on the rest of my faith as well, for it suggested that if I could be so wrong about something as important as doomsday, what was to keep me from being wrong about a whole host of other important questions such as salvation, hell, the resurrection, and other essentials of the faith? And so I started on a process of challenging—and, when necessary, abandoning—each of the many pillars of the faith I had blindly adopted years earlier. Eventually, cognitive dissonance become my constant companion, and by the time I turned forty I'd had enough and left the church to pursue other spiritual paths, none of which contained a word about the end of the world.

I had finally turned the last page of *The Late, Great Planet Earth*. In retrospect, I suppose it could be said that, in the end, it was doomsday prophecy that was largely responsible for getting me into the church, and it was doomsday prophecy that was largely responsible for convincing me to leave it as well. Curious how things work out sometimes.

The point of all this is that while the 2012 prophecies and Cayce's and Nostradamus' musings are all the rage, it is those prophecies contained in the words of the Bible that have had the most profound and lasting impact on how we view the world in which we live. As such, this would be a good point at which to examine Bible-based end-times scenarios in an effort to understand why their bleak predictions continue to hold us captive. I believe it is only in examining this element of Scripture that we might come to understand not only how doomsday prophecies came about, but also how they have grown and expanded over the centuries to become such a major element of humanity's shared fears about the future.

Chapter Eight

How End-Times Prophecy Works

I realize that Bible prophecy may not be everyone's cup of tea, but considering the impact the Bible's teachings on the subject have had on prophetic beliefs and Western civilization itself over the centuries, it is well worth taking the time to understand it better if only so we might acquire a better sense of where so many millions of Christians are coming from in their beliefs. I hope to do this in a fair and evenhanded manner, and I am content to leave it to the reader to decide if I have been successful. Additionally, while I realize that end-times prophecies are not unique to Christianity but can also be found in Jewish and Islamic writings, I'm confining my discussion to those prophecies unique to Christianity, both because I know them better and because they are more likely to be familiar to my readers.

As I wrote in some detail in the previous chapter, I was first introduced to the notion of doomsday through the writings of Hal Lindsey. His perspective, however, is by no means unique, but is based upon a uniquely Christian understanding of how the end times are to play out on planet Earth that has been around in one form or another for almost two thousand years. In a nutshell, Christianity works around the premise that just as Jesus of Nazareth died, allegedly rose from the dead, and ascended into heaven some two thousand years ago, he is also going to return from heaven to reestablish his kingdom on Earth, defeat the forces of evil, and bring all people to judgment as the final act of human history. In fact, this belief is so central to Christian thought that the very faith itself could not be considered complete until and unless this return—known as the *parousia* in Christian jargon—occurs, making it in some ways the single defining event and hope of the entire faith.

Since end-times scenarios play such a pivotal role in the faith, it's understandable, then, that the church would spend so much time and energy on the question of just when this event might occur. So important has the question become, in fact, that the church even has an official "science" called *eschatology*, dedicated entirely to the study of end-times prophecies (from the Greek *eskhatos*, meaning "last").

Not surprisingly, almost from its inception the issue has been a source of division within the church—especially in the last couple of centuries—with various schools of thought regarding just how this event is to play itself out battling each other for dominance. While all of them agree that Christ will return to judge the living and the dead and destroy Satan and his evil cohorts, what they can't seem to agree on is the sequence of events. The major sticking point, it seems, has to do with something called the *millennium* (from the Latin base word *mille*, meaning "one thousand"), a period of time mentioned in the Book of Revelation that has to do with the timing, nature, and extent of this earthly kingdom Jesus Christ is to install on the planet upon his return. In other words, Christians have been locked in debate for two thousand years as to when the millennium is to start, how long it is to last, and whether it is a literal period of time or a figurative "age" or "era." But we're getting ahead of ourselves here. First, let's take a closer look at precisely *what* the millennium is.

The millennium is the time during which Christ physically reigns on Earth, presumably from his capital in Jerusalem. It is also known as the "messianic Age" or the "church age," though most commonly it is simply referred to as the millennium. The problem comes in deciding whether this is a literal thousand-year period of time as many assume or whether it is a metaphor designed to simply denote a long period of time of uncertain duration. In other words, it may simply be a way of saying a "very long time" (much as we might use the word *forever* when complaining about the length of time we've had to wait in a restaurant for service). There is also a question of whether this period of time is to be realized in the future as an imminent "golden age" of peace and prosperity or whether it was realized in the past (or is in the midst of being realized in the present, making it simply a figure of speech or a term for the "church age"—that is, the length of time the Christian church has been in existence). To break this down, it can be said that there are generally four different perspectives Christians hold with regard to what the millennium is, when it occurs, and what, exactly, it is. Let's look at each school of thought in more detail.

Postmillennialism, Amillennialism, and the "Church Age"

The first position—and perhaps one of the oldest and most enduring within Christendom for centuries—is called *postmillennialism*. The postmillennialist position is characterized by the belief that the millennial kingdom was established upon Christ's resurrection, and that the world is being slowly transformed into a Christianized planet as a result. To the postmillennialist, the millennium *precedes* the Second Coming; it does not initiate it and, further, Christ's physical return will occur *only after* the entire world has been completely Christianized, thereby making the parousia (the Second Coming) the final act of history. In essence, postmillennialists believe that the world is, in a sense, already God's kingdom on Earth, the fruits of which we are coming to realize as the world slowly moves toward greater enlightenment and understanding. Not surprisingly, however, considering the number of wars, inquisitions, persecutions, and general mayhem the world has seen over the two thousand years since this "kingdom" was supposedly established, this has become an increasingly less popular position today than it once was. At one time, however—particularly in the late nineteenth and early twentieth centuries—it was all the rage, before the carnage of two world wars in one century gave it a black eye it has yet to recover from.

A similar position—at least in terms of the millennium being thought of as a spiritual rather than a physical kingdom—is called *amillennialism*, which means, literally, *no* millennium. The amillennialist, then, like the postmillennialist, considers the millennium to be essentially a metaphor denoting the entire period of history between Christ's resurrection and his final return, but unlike postmillennialism, the amillennialist does not see progress during this period to be indicative of Christ's imminent return. To the amillennialist the world will be subjected to various trials and tribulations just as it always has been right up to the time of Christ's return, which could happen tomorrow, next year, or 10,000 years from now. Typically, expectations concerning the reign of Christ are seen as being partially fulfilled in the past but are not to be ultimately realized until his bodily return at the end of time. In essence, this makes the kingdom of God both "now and not yet"—realized *now* in the respect that Christ effectively established his church at the moment of his resurrection and *not yet* in the respect that it will not be fully realized until later, upon his return.

What's important to remember about both these positions is that neither sets precise dates for the time of Christ's return, nor do end-times scenarios play much of a role to their proponents. Since they don't see the millennium as a literal thousand-year period nor necessarily interpret the Kingdom of God as a physical throne, doomsday scenarios are either minimized or ignored altogether. Postmillennialists may see Christ's return as being more imminent when there are positive signs of social progress or great growth within the church, but for the most part neither position looks for the end of the world to be coming anytime soon.

Premillennialism and the Rise of End-Times Theology

The third and most popular interpretation of the millennium today, as well as the one closely associated with the most popular modern end-times teachings, is called *premillennialism*, which means, literally, *before the millennium*. In other words, premillennialism, unlike postmillennialism and amillennialism, sees the millennium as a yet-to-be-realized future event rather than something that's already happened in the past or is ongoing today. Furthermore, unlike the other eschatological camps, premillennialism interprets the word *millennium* literally rather than figuratively, seeing it as a literal one-thousand-year period of time to be realized once certain historical and political events have come to pass, with Christ's physical return being the initiating event of the entire era.[1] Premillennialists also believe that this manifestation will be preceded by certain international events designed to set the stage for that return.

Lest anyone imagine that there might be complete agreement among premillennialists as to exactly how this is to be played out, the fact is that considerable debate rages among them, mostly having to do with the precise sequence of events. This has resulted in a split among premillennialists, with two main factions emerging from the rubble. The first of these is the traditional premillennialist position, which believes that the parousia could occur at any time, at which point Jesus will bind Satan and his followers and set up a thousand-year period of peace. The other position, more popularly known as *dispensationalism*, also believes this but adds that Christ's return is not going be a single event, but a twofold occurrence interspersed with a seven-year period of great tribulation during which the planet will be subjected to a time of intense destruction and distress. In other words, the dispensationalist believes that Christ will come first to rescue his church (generally defined to mean the entire body of true believers regardless of their denomination) by physically removing them from the planet in a single instance—the *rapture*—and then he will return seven years later with his previously raptured church in tow to finish off Satan and his allies in the great and horrific Battle of Armageddon. Then, and only then, does he reestablish his church on Earth, ushering in the thousand-year period of peace and justice. Each of these views, in greatly simplified form, are illustrated in the chart on the next page.

---

1. To be fair, there are some premillennialists who do interpret the millennium metaphorically, effectively leaving the duration of this golden age open-ended, but they are in the minority.

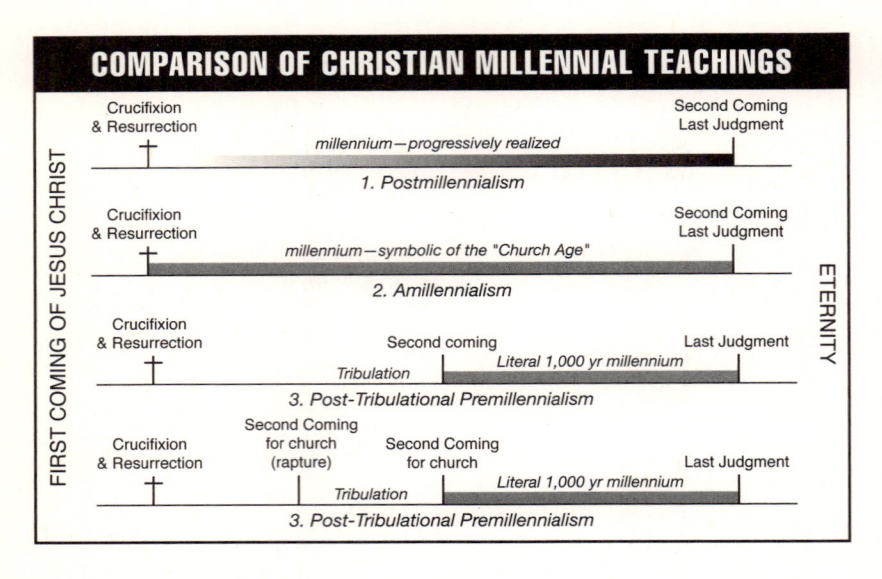

COMPARISON OF CHRISTIAN MILLENNIAL TEACHINGS

```
FIRST COMING OF JESUS CHRIST                                           ETERNITY

Crucifixion                                         Second Coming
& Resurrection                                      Last Judgment
  †         millennium—progressively realized
                    1. Postmillennialism

Crucifixion                                         Second Coming
& Resurrection                                      Last Judgment
  †         millennium—symbolic of the "Church Age"
                    2. Amillennialism

Crucifixion          Second coming                  Last Judgment
& Resurrection                  Literal 1,000 yr millennium
  †         Tribulation
               3. Post-Tribulational Premillennialism

              Second Coming
Crucifixion    for church      Second Coming
& Resurrection  (rapture)       for church          Last Judgment
  †                             Literal 1,000 yr millennium
              Tribulation
               3. Post-Tribulational Premillennialism
```

However, that's not the end of the story. While for the post- and amillennialist, Christ's return is the end of things—and technically Judgment Day for all humanity—to the premillennialist it's merely the end of the first act. Most believe that at the end of this thousand-year-long golden age of peace and prosperity—during which time Satan and his followers will be bound and therefore restrained from sowing trouble—God will inexplicably release the devil and give him free rein to once more do mischief on the planet. During this time, the devil will somehow manage to collect an army of followers (though it's not certain how he achieves this in the midst of a golden age of enlightenment), whom he will enlist to descend on Jerusalem in an effort to defeat God once and for all. Of course, their fate is already sealed, as we are told that they will be finally defeated in a great battle (the *real* Battle of Armageddon), after which the final judgment will take place and Satan and his followers—both human and demonic—will be cast into the lake of fire to suffer eternal torment.[2]

---

2. I always wondered why Satan, who presumably can read the Book of Revelation himself and can see what awaits him, doesn't just "lie low" once he's released until the thousand years are over, but then I'm not the strategist he is.

Understanding the Rapture

Particularly unique to dispensationalism is its insistence on a double or twofold parousia—the first being a "rapture" of the church at the beginning of the tribulation period to be followed by Christ's final return at the end of this period. This is done to supposedly ensure that God's "chosen ones" (all professing Christians) are spared the torment and misery of this Great Tribulation Period by being instantly and physically "snatched" from the planet (hence the term *rapture*—from the Latin *raeptius*, which in turn is a translation of the Koine Greek word *harpazo*, meaning to "snatch up" or forcibly remove). In effect, it is the escape mechanism by which all professing Christians are to be spared from having to go through the seven-year period of testing, torture, and terror God is going to unleash on the earth in an effort to convince all "unbelievers" to change their tune. It is also the mechanism that starts the seven-year clock ticking toward Christ's final return and the establishment of his thousand-year kingdom.

So popular has this idea become that Christian bookstores are filled with titles that teach the premise, but probably none have been as effective in getting this idea out to the general public as has the *Left Behind* series of novels by pastors Tim LaHaye and Jerry Jenkins. Outlining in gory detail all the horrific events that are to occur to an obstinate humanity immediately after the rapture removes all Christians from the planet, the book and its fifteen sequels have proved to be one of the best-selling Christian-themed novel series of all time. Additionally, actor Kirk Cameron, probably best known for his role on the television sitcom *Growing Pains* during the late 1980s, has starred in a trilogy of "post-rapture" movies based on the novels that have proved to be very successful with Christian moviegoers. While it's not known how effective these films are in spreading the dispensationalist perspective, they have undoubtedly contributed much to the idea that God rescues his people before turning his wrath on nonbelievers.

In any case, it is the rapture that not only makes dispensationalism—a theological position that has only been around for a couple of centuries[3]—unique, but that also makes it a major source of fear among millions of people.

So where did the idea that millions of Christians will be instantly whisked to heaven in the blink of an eye come from? Though there are a few passages in Scripture that hint at such a possibility, the clearest teaching comes from the writings of the apostle Paul, who wrote the following in his first letter to the church in Thessalonia:

> *For this we say unto you by the word of the Lord, that we which are alive and remain unto the coming of the Lord shall not prevent them which are asleep. For the Lord himself shall descend from heaven with a shout, with the voice of the archangel, and with the trump of God: and the dead in Christ shall rise first. Then we which are alive and remain shall be caught up together with them in the clouds, to meet the Lord in the air: and so shall we ever be with the Lord.* (1 Thessalonians 4:15–17)

Notice that Paul doesn't say anything about a two-part event but simply suggests that those who are alive at the time of Christ's return will not die but be instantly transported (known in some circles as being "translated") to heaven to join those Christians who died throughout history. How later interpreters turned this into the fantastic scenario seen today is a long and convoluted story, but suffice it to say the two-part Second Coming is based far more on speculation than Scripture.

Furthermore, dispensationalists can't even agree among themselves how this rapture thing works. Most believe it is the event that initiates the seven-year tribulation period (and is, in fact, the single event that makes it possible for the Antichrist to seize power); others believe the

---

3. I won't go into detail about the history of dispensationalism other than say it got its start in the writings of an Anglo-Irish evangelist John Nelson Darby about 175 years ago and was later popularized in the best-selling *Schofield's Reference Bible*.

Christians will be raptured halfway through this period;[4] while a third group—arguably the most pessimistic of the bunch—believes it will come as the final act at the end of the Tribulation (although the rationale for such timing remains a bit murky). Each group—known as pre-, mid-, and post-tribulationists—has its favorite verses to support its position and will fight tenaciously for its own interpretation.

There is also disagreement as to whether this event could happen at any moment without warning or if certain events are to precede it first. For example, many believers maintain the nations of the world must first unify their currency into a universal standard (shades of the euro?), that there must be peace in Israel (Camp David peace accord?), that there must be a one-world government (fat chance), that Herod's temple in Jerusalem must be rebuilt on its original site, and even that the Old Testament commandments concerning animal sacrifices must be reinstated. Most, however, believe either these events have already been historically realized in the distant past or will only occur during the tribulation period (and therefore *after* the rapture), meaning that the "great event" could happen at any time. In any case, it's going to be a pretty traumatic event for all concerned, especially if literally hundreds of millions of people around the world vanish instantaneously, leaving pilotless airplanes and driverless vehicles careening about.

Conclusions

Even though most Christians accept Jesus' pronouncement that " . . . of that day and that hour knoweth no man, no, not the angels which are in heaven, [nor] the Son, but [only] the Father" (Mark 13:32), many still have a tendency to "read the signs" to see if they might not be able to gauge when the rapture might occur, which is why Christians—and especially dispensationalists—tend to be, if not precise date-setters, at

---

4. Support for this position comes from the seventh chapter of Daniel (verse 25), where it says the saints will be given over to tribulation for "time, times, and half a time," which is interpreted to mean 3.5 years—or halfway through the seven years of the tribulation.

least general date-setters. In other words, while most doomsday ministers will not set a precise date for doomsday—though a few are willing to do so—almost all of them interpret every noteworthy news event, tropical storm, earthquake, and economic downturn as evidence that the "end" (or is it the beginning?) is near, making them implicitly guilty of date-setting. The fact that some of them have been at it for almost forty years demonstrates not only how good they are at convincing others, but also how almost everything can be interpreted as being a sign of the end if one believes strongly enough.

Additionally, by not setting precise dates—or being forgiven by their followers when they do and the date passes uneventfully—they ensure themselves many more years of ministry work and TV revenue. This may sound harsh, but unless Christians begin holding these prophets of doom accountable for what they say, I'm afraid the church will continue to live in a sort of pre-doomsday environment in which the future looks bleak and the only hope for a better world is for God to wipe out most of the planet and start from scratch. Personally, I can't imagine anything that would infuriate a God of love more.

Chapter Nine

More Dispensationalist Meanderings from the Bible

Some of my readers will undoubtedly accuse me of treating the dispensationalist position superficially by not examining the scriptural evidence in support of it. To satisfy those who hold to such an opinion, this chapter will examine a collection of "evidence" that premillennialism—and, most specifically, dispensationalism—points to in an effort to bolster its position that the end is nigh. Before we begin, however, I warn the reader that this chapter is going to be pretty "Bible heavy," which may prove to be tough sledding for those who have not had much to do with the traditional church. As such, some may wish to skip this chapter and move on, content with their more basic understanding of what Christians believe about the end-times and why they believe it. For those curious about the details, however, they may find this chapter of interest, if only to get a flavor of how the dispensationalist mindset works.

As I mentioned in the last chapter, what tends to get dispensationalists in so much trouble is their insistence that the timing of Christ's return and the events preceding it can be anticipated based upon a "reading" of the signs—that is, by noting various historical events they believe were prophesied thousands of years ago that are being realized now or will be realized shortly. In this respect, then, dispensationalism is built around a very complex set of preconditions that need to be met to make it all work. The problem with this is that if the events dispensationalists point to as being futuristic have been realized in the past, or if their interpretations of Scripture are erroneous on any level, the entire premise of dispensationalism collapses like the proverbial house of cards. To see how well its foundation endures, then, it is only necessary to examine a few of its "proof texts" to see how well they stand up in the light of history and logic.

Israel and the Mount Olivet Discourse

A major element common to dispensationalism is its insistence that the establishment of the modern nation of Israel in 1948 is the most historically important event of modern times. Of course, dozens of countries have declared their independence in the last hundred years with no equivalent impact upon history being evident, so why Israel's declaration should be so important when other declarations of independence are barely noticed remains a question to some people. The answer, however, is fairly simple if one looks at it from a biblical perspective: to a premillennialist, the Jews are God's chosen people, making Israel God's mechanism specifically designed to signal when Christ's return is imminent. In effect, then, according to many Bible scholars, when Prime Minister David Ben-Gurion declared Israel's independence on the evening of May 14, 1948, he supposedly started a doomsday clock ticking toward the Great Tribulation, the rapture, the unveiling of the Antichrist, and the final Battle of Armageddon that continues to run to this day, effectively making modern Israel the time meter upon which all of dispensationalism ultimately hinges.

So why do premillennialists believe Israel to be the key to the end times, and why do they consider its reestablishment in 1948 so important? It has to do with one of Jesus' most famous sermons given just a few days before his death. He and his disciples were studying Jerusalem from a hilltop known as the Mount of Olives when one of his disciples commented on how magnificent the temple Herod the Great had built fifty years earlier truly was. Jesus' response, however, was not what they were expecting: looking out upon the structure's beautiful whitewashed walls and gilded temple from their hilltop vantage point, he declared that the building would one day be destroyed so thoroughly that "... there shall not be left here one stone upon another, that shall not be thrown down" (Matthew 24:2).

Now for a Jew, such an idea was hard to fathom. The Temple was the heart of Judaism (much as the Vatican is to Roman Catholics or the Grand Mosque of Mecca is to Muslims). For it to be destroyed would be tantamount to doomsday in the Jewish world, automatically making Jesus' statement not only prophetic but even apocalyptic in nature. Not surprisingly, then, his hearers naturally assumed its destruction would be the end of time, which is why one of the disciples asked, "Tell us, when shall these things be? And what shall be the sign of thy coming, and of the end of the world?" (verse 3), at which point Jesus breaks into a lengthy sermon having to do with not only the destruction of the temple but with various events leading up to it. It's a rather lengthy discourse (but one worth reading) that he concludes with the rather obtuse statement that: "... I say unto you, This generation shall not pass, till all these things be fulfilled ..." (verse 34), implying that all the things he described would occur within the lifetimes of some of those listening to him at that moment.

Considering that a biblical generation is usually considered by most Bible scholars to be about forty years (the maximum age at which a woman could normally be expected to bear children), that suggests that everything Jesus predicted would take place within the next few decades. Assuming he made these statements around 30 CE or so (a time frame most historians are comfortable with), that means the destruction of the temple—along with all the events he had just prophesied—would take place no later than 70 CE. Imagining that most of his disciples were likely in their early twenties at the time,[1] the thought that at least a few of them might still be alive forty years later

---

1. A not unreasonable assumption. Because of Hollywood, we tend to imagine the disciples as being older men well in their thirties and forties, but it's more likely that most of them were much younger—possibly even in their teens—at the time Jesus was alive.

was not unreasonable, despite the much shorter average life expectancy at the time.

Jesus turned out to be spot on, for indeed the temple—along with the rebellious state of Israel itself—was destroyed by the armies of Rome in 70 CE, just as he had predicted, making his words not only prophetic but uncannily accurate. Of course, dispensationalists realize this as well, but have an interesting take on it. While conceding that Jesus may well have been referring to the destruction of the first-century temple, they also believe he was also referring to a second, much later event, at the same time: his Second Coming (and, presumably, another future temple). The reason for this is because of the statement he makes, in which he says:

> *And then shall appear the sign of the Son of man in heaven: and then shall all the tribes of the earth mourn, and they shall see the Son of man coming in the clouds of heaven with power and great glory. And he shall send his angels with a great sound of a trumpet, and they shall gather together his elect from the four winds, from one end of heaven to the other.* (Matthew 24:30–31)

Clearly, it's difficult to see how that statement corresponds to the Roman destruction of the temple in 70 CE, leading many to suggest that Jesus was talking about a future, as yet unrealized temple, and a future Israel as well. In effect, they see it as a double prophecy, one applying to the immediate future and the other to the far-flung future, with the second prophecy referring directly to his physical return to Earth at the end of time. Therefore, when Israel was reestablished as a sovereign state after almost two thousand years of wandering and persecution, dispensationalists jumped on the event as a dramatic fulfillment of this particular prophecy. And if the prophecy did contain a double meaning, that meant that Jesus' return must occur within the lifetime of the generation alive at the time of Israel's rebirth, and overnight the Second Coming went from being the "blessed hope" of the church to occur someday to being a very immediate event to be realized within a few decades.

So, do dispensationalists have a point? Could they be correct in their assumption that Jesus was alluding to a far-flung future event in his pronouncements? Certainly Jesus' remarks in Matthew do make it sound as though he's talking about his return to Earth—especially when he says things like "And then shall appear the sign of the Son of man in heaven: and then shall all the tribes of the earth mourn," and "... they shall see the Son of man coming in the clouds of heaven with power and great glory" (Matthew 24:30). Also, isn't it clear from the way the disciple poses his question in verse 3 that he and the other disciples—and presumably Jesus as well—are all talking about not just the destruction of Herod's temple but the end of time?

Clearly it could be interpreted as such until we compare the Mount Olivet discourse—as this particular discussion is known—contained in Matthew's gospel with the two parallel accounts of the same event as recorded in both Mark's and Luke's gospels. When we do this, we find a number of major differences that have the potential to change how we view the account in its entirety. For example, in Mark's gospel (and in Luke's closely parallel account), the disciple only asks: "Tell us, when shall these things be? And what shall be the sign when all these things shall be fulfilled?" (Mark 13:4). Note that neither Mark nor Luke say a word about the "... sign of your coming and of the end of the age ..." as recounted in Matthew.

While this may appear to be a small difference, in reality it is quite significant, for only in Matthew are we able to tie Jesus' statements to the end times. In Mark and Luke his response seems to relate more directly to the future of the gleaming temple standing before them. As such, it could be surmised that the author of Matthew's gospel (who remains unknown to this day) may have been guilty of doing a little embellishing—or, more precisely, assuming—when he penned his account.

Certainly, if true, it wouldn't be the only time the author of Matthew was guilty of taking a bit of artistic license. This is the man, after all, who records a number of dubious events the other three gospels remain entirely silent about, including—but not limited to—his historically unsubstantiated accusation that Herod ordered the slaughter of the children of Bethlehem (Matthew 2:16), that an angel told Joseph in a dream to flee to Egypt (Matthew 2:13–15), and that at the moment of Jesus' death that " . . . the veil of the temple was rent in twain from the top to the bottom; and the earth did quake, and the rocks rent; and the graves were opened; and many bodies of the saints which slept arose, and came out of the graves after his resurrection, and went into the holy city, and appeared unto many," (Matthew 27:51–53)—an event that, if true, would surely have made its way into the secular history books. As such, we can ask whether Matthew might not simply have been up to his old tricks again and had simply turned Jesus' words into some far-flung future prophecy in spite of the contextual and historical problems that entails.

What's more difficult to understand, however, is why premillennialists choose to start the forty-year generational clock (Matthew 24:34) at the founding of the modern state of Israel in 1948 rather than in the first century when Jesus made his pronouncement. Clearly, his listeners would have understood him to be talking about an event that would be happening in their lifetimes, not something that wouldn't occur for another two thousand years. If Jesus really was talking about a third temple being destroyed sometime in the twenty-first century, there is no evidence that his listeners ever "got it," making him guilty of being one of the worst communicators of the age.

I suspect, however, just the opposite to be the case. Jesus understood the political undercurrents of his era and knew it was only a matter of time before the "lid blew"—so to speak—and the Jews rebelled against their Roman masters. He also understood that the Jews had no realistic chance of defeating the mighty Roman army and that in their religious fervor they would rally to the temple to make their last stand, which the Romans would then probably destroy in their efforts to exterminate the last holdouts (which is exactly what happened forty years later). As such, his guess that the temple would be destroyed in the next few decades was not so much proof of his prophetic ability as it was evidence of his keen political insight.[2] That end-times preachers manage to miss this point consistently is not an indictment against the reliability of the Scriptures but an indictment of their own unwillingness to look at the entire story in its historical context (as well as evidence of an overblown faith in the accounts penned by some unknown writer identified only as "Matthew").

Was the Reestablishment of Israel in 1948 Prophesied?

But even if we accept that Jesus was speaking of first-century events to be realized within the lifetimes of his listeners, isn't the fact that Israel exists today significant, and, even more so, evidence that biblical prophecy is accurate?

How significant a reborn Israel has been is a matter of opinion. Certainly, it could be argued the nation's resurrection from obscurity was unprecedented, but not entirely a surprise considering the international sentiment toward a Jewish homeland in the immediate aftermath of the Holocaust. However, to say that its modern rebirth was predicted in the Bible thousands of years ago is simply not true. The fact is that nowhere in the Bible is the modern reestablishment of Israel foretold or even hinted at. Israel's modern incarnation may be important in terms of Jewish identity or to the restoration of national unity, but it has no biblical significance regardless of how badly dispensationalists want it to have such.

So where did they get the idea that it does? The scripture to which dispensationalists point in order to bolster their argument that Israel is the prophetic centerpiece of the apocalypse comes almost entirely from the thirty-seventh chapter of the Old Testament book of Ezekiel. Written some 2,500 years ago, the book records a vision shown to the

---

2. Additionally, some more liberal scholars maintain that the gospels that record the account of the temple destruction were penned some years after the event, suggesting that later chroniclers were simply guilty of a bit of postdiction—that is, of putting a realized historical event into Jesus' mouth in an effort to give the event prophetic significance and Jesus, by extension, prophetic gifts.

prophet Ezekiel concerning Israel's future—which, considering the fact that the nation had been destroyed and its inhabitants were then in exile in Babylon, was important. In this curious chapter, the prophet is shown a valley full of dried bones and is told by God that they are the bones of a dead Israel. God then commands Ezekiel to prophesy to the bones and to tell them that He will " . . . lay sinews upon you, and will bring up flesh upon you, and cover you with skin, and put breath in you, and ye shall live; and ye shall know that I am the LORD" (verse 6). Almost immediately, the prophet writes, " . . . the bones came together, bone to his bone. And when I beheld, lo, the sinews and the flesh came up upon them, and the skin covered them above: but there was no breath in them." (verses 7–8).

At this point, God instilled the breath of life into the resurrected men that stood before Him, and they became alive once more. God then explains to Ezekiel, who was uncertain as to the meaning of the vision, that the bones represented the conquered (dead) nation of Israel, then in exile in Babylon, and that their return to life was a metaphor for the return of the exiled Jews to their homeland. While some end-times scholars consider this event to refer to the reestablishment of Israel in 1948, most scholars disagree, considering this prophecy to have been fulfilled with the return of the Jews to Israel during the time of the Persian king Darius around 500 BCE (which did result in the reestablishment of Israel in its original homeland after almost a century in exile). As such, it has nothing whatsoever to do with the modern restoration of Israel, and never did.

If true, however, that's a big problem for the dispensationalist position, for if Jesus wasn't alluding to the modern restoration of the Jewish state in 1948 in his reference to the fact that he would return within a single generation (Matthew 24:34), then his prophecy is already fulfilled, throwing the dispensationalist's entire premise, along with their carefully wrought doomsday scenario, into disarray. As such, the 1948 date, while promising much initially, proved to be a great disappointment to dispensationalism, especially once the forty years (a "single generation" in biblical parlance) came and went in 1988 without incident. Dispensationalists have tried to get around this by using various devices (questioning the true length of a biblical generation, basing the forty years upon the seizure of East Jerusalem from Arab forces in 1967 rather than the founding of Israel in 1948, and so on), but in every case they appear to come up short, making dispensationalism a precarious position to maintain with anything approaching intellectual integrity.

Finally, it can be argued that this emphasis upon Israel as the centerpiece of prophecy has had a deleterious effect on the church in general, both in its credibility (how many people have either left the church or simply stopped believing it after so many decades of failed prophesying?), and in Christendom's relationship to the Muslim world, which increasingly—and often rightfully—perceives Christianity as an ardently pro-Israel, anti-Arab lobby indifferent to the loss of Arab land to Israel. In doing so, has the church inadvertently alienated nearly a billion people because of the growing perception among Muslims that Christianity is simply a mouthpiece for Zionist nationalism?

Other Failed Elements

The establishment of Israel is not the only part of doomsday teaching that fails to pass the test. Another major element of dispensationalism is the idea that as part of the tribulation period (perhaps even the opening event after the rapture), Israel is going to be invaded by an army "from the north"—usually interpreted to be a coalition of eastern European nations led by a resurgent Russia, which is the event that is to make it possible for the mysterious figure known only as the Antichrist (more on him in the next chapter) to emerge onto the world stage.

The impetus for this idea is largely lifted from the same aforementioned book of Ezekiel—this time the thirty-eighth chapter—in which God talks about someone named "Gog and Magog" (along with a host of allies) sweeping out of the north to invade Israel. Generally, dispensationalists interpret these names to be ancient designations for various modern nations destined to one day invade Israel. In other words, some Bible commentators speculate that Gog and Magog are simply metaphors for the old Soviet Union, and the other "tribes" are monikers for her eastern block and Middle Eastern/African allies. However, most modern scholars disagree with this interpretation; while not entirely certain precisely who these tribes represent due to the fact that countries and regions were called by many different names over the centuries, they are fairly sure they represent *ancient* invaders of Israel and not modern Russia. For example, Gog and Magog could be a reference to the Seleucid Greeks, who invaded Israel in the fourth century BCE and were to rule the country for the next two hundred years.

Unfortunately, without a good working knowledge of the region—a land that saw numerous invasions and battles throughout its long and bloody history—and considering the geopolitical climate of the last half of the twentieth century, the idea that the Soviet Union and her allies might have tried to invade the Middle East seemed quite reasonable—at least, it did until Communism fell and Russia became comparatively docile as far as world empires go. Still, the notion that Israel will one day be invaded by a confederacy of nations "from the north" remains a part of end-times scenarios to this day, though with less urgency than it did twenty years ago.

The Reconstituted Roman Empire

Another idea popular with dispensationalists is that during the Great Tribulation period the ancient Roman Empire will be revived to serve briefly as the world's capital and as a base for the Antichrist. The idea finds most of its support in the Old Testament book of Daniel, where in chapter 7 the prophet describes a dream he had in which he watches a series of unusual creatures emerge from the sea: the first being like a lion with the wings of an eagle, the second like a bear with three ribs in its mouth, the third a four-headed leopard with four wings, and the fourth—the most frightening and terrible of all—a creature with ten horns, one of which " . . . had eyes like the eyes of a man and a mouth that spoke boastfully" (verse 8). While end-times proponents accept the first three animals as being symbolic of earlier empires in history (the Babylonian, Persian, and Greek Empires, to be precise), most interpret the fourth to be a metaphor for a yet to be realized empire greater and more terrible than all the others that preceded it. In fact, dispensationalists maintain that this beast is the symbol for the kingdom the Antichrist is to establish in the first days of the Great Tribulation. In effect, they interpret the passage as a metaphor for a confederacy of European nations, led and controlled completely by a single "super dictator."

With the formal establishment of the European Economic Community in 1957 (the Treaty of Rome), this interpretation quickly gained support, especially in the 1973 when the EEC appeared to be on the threshold of acquiring its tenth member nation (the "tenth horn" of Daniel's vision). However, things began quickly going awry when Norway—which would have been the tenth member—rejected membership, leaving the EEC at nine members until the 1980s, when three more countries—Spain, Portugal, and Greece—joined, thereby throwing the numbers off. Today the European Union (the EEC's new name after the Maastricht Treaty went into effect in 1993) has twenty-seven members, truly challenging the ancient prophecy's accounting acumen. Furthermore, it appears to be nothing like the terrible and powerful beast of Daniel's vision, but remains largely an economic union rather than a true military/political alliance.

While it remains to be seen how extensive and effective this union may ultimately prove to be, it appears unlikely to ever live up to Daniel's aspirations nor is it likely to be open to installing the single world leader so dear to dispensationalist teachings. This doesn't prevent many from continuing to hold on to the hope that one day it may indeed become the brutal empire of prophecy, however, regardless of how unlikely and even misguided that hope might be.

So, if not a powerful federation of European nations, then what does this last beast truly represent? It's not entirely certain, but some scholars feel it could find its fulfillment—at least in part—in the emergence of the Roman Empire, a political/military entity that truly was the greatest and most terrible of them all—at least in the context of that era. What the ten horns represent may be a source for some debate, of course, but that Daniel was simply using vivid imagery to outline the emergence of each of the great empires of his age seems considerably more reasonable an interpretation than does the prospect that he was foreseeing the dawn of a modern European empire more than twenty centuries in the future.[3]

The Third-Temple Controversy

The Book of Daniel is also responsible for another major element of dispensationalist teachings: which is that the rebuilding of the ancient temple, the reestablishment of the old Mosaic laws, and the return to the ancient Jewish rituals and animal sacrifices are supposed to be realized both before Christ can return and the Antichrist can be revealed. Writing in the ninth chapter, the prophet says:

> Know therefore and understand, that from the going forth of the commandment to restore and to build Jerusalem unto the Messiah the Prince shall be seven weeks, and threescore and two weeks: the street shall be built again, and the wall, even in troublous times. And after threescore and two weeks shall Messiah be cut off, but not for himself: and the people of the prince that shall come shall destroy the city and the sanctuary; and the end thereof shall be with a flood, and unto the end of the war desolations are determined. And he shall confirm the covenant with many for one week: and in the midst of the week he shall cause

---

3. As a side note, it is interesting that Daniel managed to foresee the emergence of the Roman Empire centuries before it was to arise, implying that Daniel's prophetic gifts—at least in this instance—may have been legitimate. However, this depends upon when the Book of Daniel was actually written. Tradition has maintained that the book was written in the sixth century BCE (which would have been long before the city state of Rome was a major political entity). However, many modern scholars—both Jewish and Christian—suggest it may have been a much later work, possibly having been written as late as the second century BCE, at which time Rome was already a fairly well established military/political force in Europe, thereby rendering Daniel's prophecy less remarkable.

> *the sacrifice and the oblation to cease, and for the overspreading of abominations he shall make it desolate, even until the consummation, and that determined shall be poured upon the desolate.* (Daniel 9:25–27)

While I grant you that these versus are somewhat confusing, they appear to have something to do with the idea that the magnificent temple of Solomon, destroyed by the Babylonians in the seventh century BCE, was to be rebuilt over a period of seventy "weeks," at which point the rituals and sacrifices of the old Judaic religion would then be reestablished. At the midpoint of the final or seventieth "week," however, someone will set up an "abomination that causes desolation," which will be destroyed by God.

The "weeks" are reinterpreted by dispensationalists as meaning years, making this period equal to something like 490 years (depending upon which "years" one makes use of—the Hebrew year being 360 days in length as opposed to our 365). The "Anointed One" is also interpreted by modern scholars to be another term for the Messiah, although this also is not entirely clear. The identity of the other person—the one who sets up the "abomination that causes desolation" in the temple—is equally uncertain, but is considered by dispensationalists to be the Antichrist of end-times fame. The assumption, then, is that this temple is some future construction that's yet to be realized, and one that is absolutely essential to Christ's return.

The problem is that this prophecy has a historic realization. First, if Daniel wrote his prophecy sometime around 500 BCE as tradition maintains (although modern scholars frequently assign a much later date to this work), he could easily have been speaking of the second rebuilding of the temple, which occurred shortly after the Jews returned from exile in Babylon. What's unclear is whether this second temple was a genuine building or simply represented the Jew's former system of worship, or whether it represents the magnificent temple built by Herod the Great in 19 BCE in an effort to incur the good will of his subjects (this would have been the temple in existence in Jesus' day). If we interpret Daniel's seventy weeks as being a metaphor for seventy weeks of years (490 years), it would seem to imply it is about Herod's temple, since from the time the Jews returned to Israel from exile in 500 BCE to the time that Herod finished his great temple around 19 BCE would have been approximately 490 years, largely in accordance with Daniel's prophecy. Of course, if this is what Daniel was talking about, that truly would be an impressive feat of prognostication (regardless of the century in which Daniel's book was actually written).

However, Herod's temple is not what dispensationalists are looking for. They want Daniel's prophecy to apply to an as yet unrealized future temple, and they get around the problem by deciding that Daniel's seventy-week clock stopped on the sixty-ninth week and that the final or seventieth week was not to be realized until the end times, thereby allowing for both Herod's temple to be built as well as surmising the construction of yet another temple in the future. It's not entirely clear how they get away with this, as Daniel nowhere implies this time stoppage, but stop it they do, and with great effect.

But doesn't the fact that Daniel talks about the "Anointed One" and some great evil person who "cuts off" the Anointed One imply that this temple exists far in the future? Not necessarily, for there are any number of biblical scholars who maintain that the Anointed One and the one who opposes him are historically realized in the personage of two rather obscure men from ancient history: the Jewish religious/military leader Judas Maccabees and the Seleucid ruler Antiochus Epiphanes.

The first part of solving the puzzle of the ninth chapter of Daniel is to understand the term *Anointed One* as synonymous with the Jewish term *Messiah*. Unfortunately, the word *Messiah* is understood by Christians to mean the personification of God and as such to be a reference to Jesus Christ (Christ being the Greek equivalent of Messiah). However, the term would have been understood quite differently by Jews of the era. For them, the Messiah was not some deity sent to cancel the "sin debt" between God and His fallen creation and bring spiritual salvation to all who call upon his name, but a type of warrior king specifically anointed by God to restore the kingdom of

Israel to its former glory. As such, when most Jews think of the Messiah, they are looking for a liberator or, more precisely, a military commander along the lines of Joshua, Gideon, or King David.

While little known to most people today, Judas Maccabees, a Jewish rebel commander who led a successful revolt against the Seleucid Empire in 167 BCE, was precisely that sort of person. He is acclaimed as one of the greatest warriors in Jewish history; not only did he and his brothers, fighting against overwhelming odds, manage to oust the Seleucids from Israel, but in doing so he was able to establish an independent Israel that was to stand for over a century. Therefore, from the perspective of Jews living in the second century BCE, Judas Maccabees *was* the Messiah—the "anointed one" of prophecy—and, as such, the fulfillment of Daniel's prophecy.

And as for his opponent, he too is found in the person of the Seleucid leader Antiochus Epiphanes, whose heavy-handed style in ruling Israel and efforts at Hellenizing the population by setting up pagan idols for worship—often within the Jews' own temples and holy places (the abomination of desolation?)—instigated the very rebellion that eventually brought him down. Though he was to die of disease while on a military campaign far from Israel in 164 BCE, his death made victory for the Maccabeans possible and, if perceived within the context of Daniel's vague prophecy, could be seen as a historical fulfillment of that prophecy.

One problem with this interpretation is that Daniel's prophecy seems to suggest that the Anointed One and the abomination that causes desolation are to meet at the end of the seventy weeks, whereas Judas Maccabees and Antochus Epiphanes are figures that predate the building of Herod's temple by over a century. This would seem to eliminate them as being the subject of this prophecy, at least if we are to remain consistent to Daniel's seventy-week (490-year) timetable.

However, there is another possible explanation, and it would have more to do with Jesus of Nazareth and another famous figure from ancient history, Pontius Pilate. According to some scholars, Daniel's reference to the "Anointed One" really may have been a reference to Jesus Christ, while the "abomination that causes desolation" was a moniker for Pilate, who early in his reign made the mistake of having his men post their standards on the temple grounds, which would have been considered—in that the pagan Roman standards were considered idolatrous in nature—an insult to the sanctity of the temple. This resulted in considerable unrest among the Jews and even threats of insurrection, which eventually forced Pilate to relent and have the standards removed in an effort to restore order. As such, the "abomination that causes desolation" may have been a code for the actions of a man who did, indeed, place an "abomination" in the "House of God" while his countrymen were to eventually be responsible for "causing desolation" when the city was raised in 70 CE by the Roman legions under Titus.

Whichever the case may be, if either is the true interpretation of Daniel's prophecy, both do tremendous damage to the entire doomsday scenario taught by so many today, further undercutting the dispensationalist position even more. Whether it renders it void remains to be seen, but considering the fervency with which so many hold to their end-times scenarios, I doubt if even the most self-evident events of history are going to change anyone's mind.

Conclusions

While such interpretive faux pas may be mostly harmless to the majority of Bible expositors, in the hands of premillennialists they are truly dangerous. In contriving an entire doomsday scenario by essentially pulling passages of Scripture out of context, ignoring the likely historical realization of many passages, and interpreting allegorical imagery as literal, they have had the effect of painting a very dark picture of the future and potentially leading many people into fear and discouragement. While admittedly such a scenario is more exciting and dramatic than are the historical facts of the ancient past, that does not give end-times preachers the right to frighten their flocks unnecessarily by teaching that the Bible predicts that a dark and violent world awaits them and their children. Ministers have an obligation to the truth, no matter how much the truth might conflict with their own prophetic affections. To ignore these facts is not only intellectually dishonest, but it does a disservice to the God of love.

However, the biggest thing that most dispensationalists miss, and I think what does the most damage to the church in general, is the real meaning behind what they're saying with their end-times teachings. If the events described by many end-times ministers today are true and accurate representations of what is about to befall the planet and its population in the near future, consider that what they tell us about God. Doomsday preachers tell us that Jesus is returning to Earth to rescue "his" Christians so they might be spared God's much deserved wrath, leaving the rest of the planet—about six billion people (give or take a few hundred million)—to fend for themselves. They also teach us that God is intent upon inflicting seven years of horrible plagues upon the planet, all designed, presumedly, to get humanity to repent of its wickedness. Then the very same Jesus who was willing to give his life on the cross so humanity might be saved will return to the planet that crucified him in order to wreak tremendous destruction upon billions of its inhabitants and establish his benevolent empire of peace upon their dead bodies.

While many who preach this have no trouble with such a premise, it hardly seems consistent with the idea of a loving God or with the image of Jesus portrayed in the Bible. Instead, what we see is an angry god and, even more perplexing, a wrathful and judgmental messiah completely at variance with the messenger of love and compassion demonstrated throughout the gospels. Even the most steadfast Christians cannot maintain that they like these sides of their deity; even if they believe that such actions may be necessary, no person who understands what compassion is could possibly accept these end-times scenarios as anything but brutal. Furthermore, they lack any semblance of logic. The idea that God would find it necessary to kill billions of his creatures and reduce much of the planet to cinders in order to convince humans to repent seems entirely counterproductive. Yes, one can use fear and threats to acquire allies, but such allies are superficial at best and will bolt at the first opportunity. Can conversions achieved through torture be any more genuine or heartfelt?

Such heavy-handed tactics hardly seem worthy of an omnipotent god who surely must have other, less destructive methods of enticing His creation to Himself.

Finally, if taken to their logical conclusions, the end-times scenarios outlined by Hal Lindsey, Tim LaHaye, Jack Van Impe, and a host of like-minded dispensationalists actually make God the author of evil, as evidenced by the fact that it is God who first gives Satan free rein to ravage the earth and then releases him again at the end of the millennium to once more wreak havoc upon the planet. According to the much revered Book of Revelation, it is *God* Himself who clearly has complete control of this rather nefarious entity we know as Satan. Thereby, by releasing him back upon the world, where he can create even more destruction and entice billions into following him to their own eternal damnation, could not the argument be made that Satan is little more than God's puppet—thereby making the Creator the true author of death? It's a frightening thing to consider, but remember: it is God's own angel who opens the seven scrolls that result in all the destruction poured out upon the planet; it is His own will that both binds and loosens Satan; and it is God's own willingness to allow the bulk of His creation to be damned that we need to recognize.

Clearly, if true, God is the author of the destruction outlined in the Book of Revelation—not Satan, not the Antichrist, not wicked humanity—but God Himself. This is the real message of the Christian doomsday preachers, whether they are able or willing to acknowledge it. They claim God to be good, but if their end-times beliefs are correct, He is in reality far more evil than is the very serpent He claims to be attempting to defeat. Indeed, it could be that *Satan may actually be the force for good and God the destroyer of that good*, turning our understanding of God's love entirely on its head.

But such, of course, is nonsense. God is love, and He exists in a world of light and compassion. It is some of his more misguided and angry followers who have inadvertently made Him into the villain of Revelation; and it is they who, through their own presumptuousness and mean-spiritedness, continue to keep the world from exploring His love.

Jesus may one day return to Earth as many of his followers claim, but I believe that if he does it won't be as a conquering hero intent on slaying his enemies but as a child of compassion longing to forgive a world engulfed in darkness. God is the ultimate healer, and who could possibly understand humanity better than the very deity who lived among us for a time and died so that we might better understand what real love looks like?

Chapter Ten

The Antichrist

No study of biblically based end-times scenarios would be complete without examining one of the most famous characters inherent to the entire premise, the mysterious and frightening figure of biblical and Hollywood fame known simply as "the Antichrist."

For those unfamiliar with this fellow, the Antichrist—sometimes referred to as "the Beast of Revelation" and the "abomination that causes desolation"—is a political/religious figure whose arrival during the final stages of the end times marks the beginning of the end of planet Earth—or, at least, Earth as it presently exists. Allegedly a charismatic figure of great guile and evil intent, he is supposedly granted supernatural powers by Satan himself (it's not clear whether this is a result of being directly possessed by the devil or whether it's due to powers granted to him through his alliance with the prince of darkness), which he uses to deceive the public and acquire unsurpassed power over the world—or at least those parts under his control.

While the term *Antichrist* is used frequently in doomsday prophecy, it's interesting that this figure is mentioned by name only a couple of times in Scripture, and even then only in two of the lesser known books of the Bible. In the brief epistles of 1 and 2 John, the author—thought by some to be the actual disciple John but considered by most scholars to be an unknown author—writes:

> *Little children, it is the last time: and as ye have heard that antichrist shall come, even now are there many antichrists; whereby we know that it is the last time.* (1 John 2:18)

Curiously, this passage of Scripture, written almost two thousand years ago, seems to imply that the Antichrist is not some future figure yet to appear, but multiple characters who have already come. Furthermore, just a few versus later, he tells us that these antichrists are not political or religious figures, but simply metaphors for anyone who denies the deity of Christ:

> *Who is a liar but he that denieth that Jesus is the Christ? He is antichrist, that denieth the Father and the Son."* (1 John 2:22)

> *. . . every spirit that confesseth not that Jesus Christ is come in the flesh is not of God: and this is that spirit of antichrist, whereof ye have heard that it should come; and even now already is it in the world.* (1 John 4:3)

> *For many deceivers are entered into the world, who confess not that Jesus Christ is come in the flesh. This is a deceiver and an antichrist.* (2 John 1:7)

Clearly, then, the case that the Antichrist is some super-powerful world leader is not supported by Christian scripture. In every case were the term is used, it appears to refer either to a nonbeliever or, in some instances, an attitude of disbelief. Of course, since most of the world's population does not believe Jesus of Nazareth to be divine, that would suggest that the world is populated with billions of antichrists, which seems a bit harsh. More likely, the writer of these passages had in mind those *inside* the church who were spreading what was, in his opinion, a false gospel that denied the divinity or resurrection of Christ. Who these individuals might have been is never specifically revealed, though it does appear they were contemporaries of the writer and not some future "super dictator."

However, there is another figure named in Scripture who, while appearing under a different title, does appear to fit the doomsday scenarios better. This figure is known simply as "the beast" and is first mentioned in the eleventh chapter of the mysterious Book of Revelation:

> *And when they shall have finished their testimony, the beast that ascendeth out of the bottomless pit shall make war against them, and shall overcome them, and kill them.* (Revelation 11:7)

Now, not much is known about this figure at this point, but the author of Revelation (also traditionally thought to have been the actual disciple of Jesus but more recently considered to be another man by the same name) gives us more detail in the thirteenth chapter, when he writes:

> *And I stood upon the sand of the sea, and saw a beast rise up out of the sea, having seven heads and ten horns, and upon his horns ten crowns, and upon his heads the name of blasphemy. And the beast which I saw was like unto a leopard, and his feet were as the feet of a bear, and his mouth as the mouth of a lion: and the dragon gave him his power, and his seat, and great authority. And I saw one of his heads as it were wounded to death; and his deadly wound was healed: and all the world wondered after the beast. And they worshipped the dragon which gave power unto the beast: and they worshipped the beast, saying, Who is like unto the beast? Who is able to make war with him?*
>
> *And there was given unto him a mouth speaking great things and blasphemies; and power was given unto him to continue forty and two months. And he opened his mouth in blasphemy against God, to blaspheme his name, and his tabernacle, and them that dwell in heaven. And it was given unto him to make war with the saints, and to overcome them: and power was given him over all kindreds, and tongues, and nations. And all that dwell upon the earth shall worship him, whose names are not written in the book of life of the Lamb slain from the foundation of the world.* (Revelation 13:1–8)

This imagery does, indeed, seem to suggest a literal political/religious figure, but who could it be? The author of Revelation gives us a clue:

> *And I beheld another beast coming up out of the earth; and he had two horns like a lamb, and he spake as a dragon. And he exerciseth all the power of the first beast before him, and causeth the earth and them which dwell therein to worship the first beast, whose deadly wound was healed. And he doeth great wonders, so that he maketh fire come down from heaven on the earth in the sight of men, and deceiveth them that dwell on the earth by the means of those miracles which he had power to do in the sight of the beast; saying to them that dwell on the earth, that they should make an image to the beast, which had the wound by a sword, and did live. And he had power to give life unto the image of the beast, that the image of the beast should both speak, and cause that as many as would not worship the image of the beast should be killed. And he causeth all, both small and great, rich and poor, free and bond, to receive a mark in their right hand, or in their foreheads: And that no man might buy or sell, save he that had the mark, or the name of the beast, or the number of his name. Here is wisdom. Let him that hath understanding count the number of the beast: for it is the number of a man; and his number is Six hundred threescore and six.* (Revelation 13:11–18)

Now, however, is where things start to get confusing; the writer of Revelation seems to suggest that there would be a second beast, this one subservient to the first. This figure is later referred to as the False Prophet (Revelation 16:13, 19:20, and 20:10), and he/it appears to be a sort of lieutenant to the main beast, which has led some to speculate that he is a purely religious figure, whereas the beast is mostly political in nature. In either case, they appear to be a powerful and dangerous combination.

Naming the Antichrist

So who could this figure (or figures) be and, even more important, is he someone that lived in the historic past or is he alive on the planet today (or will be in the near future), as so many doomsday preachers maintain?

Unfortunately, this uncertainty has led some to play what could only be described as the "identify the Antichrist" game, resulting in almost every major political and military leader throughout history being identified as the Antichrist at some point, much to the detriment and occasional bemusement of society in general. Hitler, Napoleon, Saddam Hussein, and Joseph Stalin have all been suggested as candidates to fill the bill, only to have such talk die down after their deaths. Italian dictator Benito Mussolini was a popular choice in the 1930s because of his Italian heritage (and the belief that the Antichrist would come out of Rome), but he eventually fell far short of Antichrist stature and was soon forgotten. Even U.S. president Ronald Reagan was suggested due to the fact that each of his three names—Ronald Wilson Reagan—were six letters each, denoting the mark of the beast.

But perhaps no figure has been more popularly and universally tagged with the moniker of antichrist over the last few centuries as has each succeeding pope of Rome. After the Protestant Reformation of the sixteenth century, not a single pope was spared being named as the Antichrist by nearly every Protestant reformer throughout the Renaissance, and they continue to be a popular target to many clergymen even today. Additionally, as many of the popes proved themselves to have been men of less than sterling character, this often appeared to be a safe bet—especially when one considers that it was from the throne of Saint Peter that such crimes as the Inquisition and witch burnings were instituted. Fortunately, however, none of the various men who have held the papacy managed to rise to the stellar heights of power described in the mysterious Book of Revelation and, as each died without ushering in the Second Coming of Christ, their names had to be removed from the roster of potential antichrists. Still, there are those who consider the pope of Rome to be the leading candidate to fill the vacancy.

Of course, simply identifying this character is only part of the problem. The idea that any one man—no matter how charismatic, charming, or otherwise persuasive—could get the various nations of the earth, with all their different languages, religions, and social structures, to follow him is difficult to imagine. Even the most charismatic leaders in history have had little success getting others outside their own borders to follow, making the idea that any single figure—supposedly emerging from Europe (according to some Bible prognosticators)—could rally the entire planet to him extraordinarily naïve. Such would simply fly in the face of everything we know about human nature and cultural dynamics.

But if the Antichrist—or the Beast of Revelation—is not a real person to emerge at some future date, doesn't that make the entire belief in such a figure nonsense?

Not necessarily. Proponents of the future Antichrist scenario all make the same mistake of assuming that the prophecy points to a figure from the far-flung future. However, what if the author of Revelation was not speaking of someone who was yet to come, but rather of someone who was *already alive* at the time he wrote his words? In other words, instead of the Beast of Revelation being a twenty-first century figure, what if he was actually someone from the first century CE?

But who could this man have been, and, furthermore, why didn't the author of Revelation simply identify him by name? The key is found in the thirteenth chapter of Revelation, where the author identifies the person with the obtuse statement:

> *Here is wisdom. Let him that hath understanding count the number of the beast: for it is the number of a man; and his number is Six hundred threescore and six.* (Revelation 13:18)

666 and the Mark of the Beast

So what does this number mean and, even more to the point, why use a number at all?

That part, at least, is easy. If the writer of Revelation directly divulged the name of the person he was talking about—and assuming that person was someone of considerable power and authority—the writer would have been guilty of sedition and could have easily found himself being hunted down and killed for making his treasonous statements. As such, to protect both himself and his readers from persecution, he had to word it in such a way that his audience would be able to "crack the code" while those outside the church would remain clueless. Since letters in the ancient alphabets often had numerical equivalents, if one wanted to identify a person using only the numerical value of the letters in his name, this would be an especially clever way to do it.

Okay, so who was the writer attempting to name? To uncover that, it is first necessary to recognize the political context at the time the book was written. Tradition has maintained that Revelation was written very late in the first century—around 96 CE, according to some scholars—but recent evidence has demonstrated it to have been an earlier text that may have been written as early as 65–70 CE.

Now, it's important to realize that if it was written earlier, it was probably penned during the time of Nero's most aggressive persecutions of the tiny Jewish sect known as Christians. As such, many Christians—and remember, at this point much of the Christian church was made up of Jews—lived in fear for their lives and were often on the run from the Roman authorities. And it was in such an environment that some Jewish Christian—probably writing in anonymity for his own protection—penned his classic tome in an effort to encourage the early Christians by illustrating how their enemies were ultimately going to be overthrown and judged by God. In essence, Revelation is a "take heart, for soon God will deal with your persecutors" type of work, designed to help the church endure its then-current persecution.

The "beast," then, was simply the writer's way of pointing to the source behind the persecutions—and, as such, the one person especially set apart for destruction by God—without naming him outright. Taking all that into account, and considering that the author needed to keep the identity of his enemy a secret, only one person makes sense: the Antichrist/Beast of Revelation is none other than one of the most feared and hated of all the Roman Emperors—the mad emperor Nero.

The Case for Nero

How do I figure?

First, Nero was the first Roman emperor to formally persecute Christians purely for their faith alone. Blaming them for a fire that ravaged Rome in 64 CE (a fire that some have argued he orchestrated in order to clear choice property for new construction), Nero, during the next three years, reportedly ordered thousands of followers of the burgeoning Christian faith to be crucified, torn apart by wild beasts in the Coliseum for the amusement of the crowds, or burned to death. If true, that makes the rationale for coding Nero's name all the more fitting.

As I noted earlier, to identify any particular Emperor as "the beast" outright would have been politically unwise. Just being found in possession of writings that attacked a sitting emperor and implying his downfall and destruction could have resulted in arrest and, very likely, death, making it necessary that any such writings be couched in metaphor to avoid bringing the sword of Roman justice down upon either the writer or his readers. However, it would still have to be written in such a way that his readers would be able to easily calculate the "number of the Beast" within the context of their era. Therefore, assigning each of the letters in the person's name a numerical value and adding them together would, if done correctly, reveal the name of the emperor in question.

Fortunately, this would not be difficult to do, for in that era letters often had numerical equivalents, much as Roman numerals still do today (I=1, V=5, X=10, and so forth). And, just like Latin, each letter in the Hebrew alphabet also had a numeric value, with the first nine letters representing 1 through 9, the tenth letter representing 10, the nineteenth letter representing 100, and so on.

Since the writer of Revelation was likely a Jew, he undoubtedly used the numerical equivalents contained within the Hebrew alphabet to name the emperor which, though he wrote in Greek, would still have been decipherable to his largely Jewish readership. Additionally, he would have worked the name from its Hebrew transliteration, which is *Neron Kesar* (Nero Caesar) or simply *nrwn qsr* (since Hebrew has no vowels). Therefore, when we take the letters of Nero's name and spell them in Hebrew, we get the following numeric values: n=50, r=200, w=6, n=50, q=100, s=60, r=200 = 666.[1] This would appear to be the clearest evidence to date, then, that the man the author of Revelation was naming was the very emperor who had been the first and most virulent of the anti-Christian (antichrist) emperors: Nero himself.

Unfortunately, modern Bible doomsday preachers tend to overlook this quite adequate answer as to who, exactly, the writer of Revelation was referring to with his obtuse reference to the number 666. The problem has been even further exacerbated by the fact that some end-times enthusiasts maintain the number of the beast to be three sixes, whereas Revelation actually maintains the actual number to be six hundred and sixty-six—a small point that has created great confusion.

---

1. Some Greek New Testament manuscripts read 616 instead of 666, but this may be a result of the Book of Revelation being translated into Latin. Possibly a Latin copyist might have thought that 666 was an error because Nero Caesar did not add up to that number when transliterated into Latin, but instead equaled 616. He may then have changed the number to conform to the Latin rendering since it was generally accepted that Nero was the Beast. In either case, a Hebrew transliteration nets 666, while a Latin spelling nets 616.

The Mark of the Beast Explained

But what of the passage that people would have to have this number embedded on their arm or on their forehead in order to buy and sell?[2] What does that have to do with the number being the numerical equivalent to Nero's name when transliterated into Hebrew? Moreover, doesn't it suggest the establishment of some sort of economic system that would seem far removed from Nero's era?

I admit that this element of Christian scripture was a source of considerable confusion to me for many years. It does appear to argue against the prospect of the number simply being a way of subtly identifying the emperor Nero, thereby giving the end-times proponents much needed ammunition. Then one day I inadvertently stumbled across a rather obscure verse of Scripture that almost knocked me over. The passage comes from one of the oldest books in the Bible, the book of Exodus. In chapter 13, we read:

> And it shall be when the LORD shall bring thee into the land of the Canaanites, as he swore unto thee and to thy fathers, and shall give it thee, that thou shall set apart unto the LORD all that openeth the matrix, and every firstling that cometh of a beast which thou hast; the males shall be the LORD's . . . therefore I sacrifice to the LORD all that openeth the matrix, being males; but all the firstborn of my children I redeem. And it shall be for a token upon thine hand, and for frontlets between thine eyes:[3] for

---

2. Some take this quite literally and believe people will have the actual numerals 666 permanently tattooed on their forehead or arm, while others believe it will be a type of individualized code that may only be readable to a special scanner. Some have even suggested that it could be a microchip that might be embedded in one's forehead or arm that will contain all of a person's personal data (as well as perhaps serve as a type of tracking device). In any case, one would be hard-pressed to survive without this code, while possessing it would be tantamount to committing spiritual suicide.

3. Modern translations render this as: "And it will be like a sign on your hand and a symbol on your forehead . . . ," which I think is clearer.

by strength of hand the LORD brought us forth out of Egypt.
(Exodus 13:11–16)

But what does this verse in Exodus mean? It's not entirely clear, but it seems to imply that remembering God's works by sacrificing the first-born animal and dedicating the first-born male of every household was the equivalent of receiving a "mark on the hand and a symbol on the forehead." Clearly, this does not appear to be a literal mark but was intended to be understood as a metaphor. In this case I might hazard that the hand represents work and the forehead represents the mind. Therefore, a reasonable interpretation of this passage might be that the Israelites were to remember God through the labor of their hands as well as keep Him in their thoughts at all times—the just-mentioned dedication ceremony being one way to remind them to do so.

Now notice how similar Exodus 13:16 is to Revelation 13:16, "And he causeth all, both small and great, rich and poor, free and bond, to receive a mark in their right hand, or in their foreheads." More than just a coincidence?[4] Could the similar verse in Revelation mean much the same thing, then, but this time as a symbol representing God's enemies? In other words, instead of receiving a literal mark on the hand or the forehead, could it simply be a metaphor symbolizing those who supported and even encouraged Nero's persecution of the Christians?

Certainly, one of the aspects of that persecution was that it was often difficult for believers to buy or sell because of their faith—something that wouldn't have been a problem had they simply repudiated their newfound faith and supported the status quo. Might not Revelation 13:16, then, be nothing more than a metaphorical means of identifying those who resisted "God's" church by supporting the hated Roman emperor economically (the works of their hands) and politically (their conscience as symbolized by the forehead)?

---

4. And notice the remarkable coincidence that both chapters and verses—13:16—are the same despite the two books being written hundreds of years apart.

Conclusions

If Nero truly was the target of John the Evangelist's warnings in Revelation, it spells serious trouble to many end-times proponents. First, it would suggest that the much-anticipated "mark of the beast," so popular within end-times scenarios, may be nothing more than a metaphorical indictment against first-century Romans who supported Nero in his persecution of the fledgling Christian church, and not some future hi-tech means of economic control. Second, if the thirteenth chapter of Revelation is a thinly veiled reference to a long dead Roman emperor, it implies that the biggest element of dispensationalism and the subject for many doomsday scenarios (and the basic plot points for a number of best-selling novels) is entirely bogus and has been from the beginning, which would obviously have profound implications for nearly all doomsday scenarios today. And, finally, by naming the beast of Revelation as Nero, it guts the entire antichrist/end-times scenario so popular among many Christians and, even worse, it suggests the Book of Revelation itself—with all its bizarre imagery and obscure symbolism—may have been already realized, leaving the premillennialist no stage upon which to build his doomsday scenario.

I can't imagine anything that would so challenge the belief of millions of people in the inerrancy of Scripture or, at a minimum, their confidence in their ability to interpret it correctly, but then again, perhaps that's the whole point. Perhaps God doesn't want us to know the future, and modern eschatology is His way of demonstrating how futile is the attempt to do so. Or maybe He's telling us to just live our life as best we can and be content in the fact that one day each of us will learn what's to happen next by leaving this earthly plain and encountering Him personally. I always suspect He does better one on one with His creation than He does through the pages of two-thousand-year-old literature written by men who have been dead longer than most nations have been on Earth. Just an opinion, I grant you, but one perhaps worth considering.

Chapter Eleven

Real Doomsday Scenarios

Having looked at doomsday prophecies of the past and the future in some detail (and usually finding them wanting in either accuracy or plausibility), it would be easy to dismiss the entire prospect that we are on the verge of destruction as just so much superstition and move on, but that would be a mistake. Simply because doomsday prophecies have traditionally proved themselves to be such poor indicators of future events doesn't mean that Earth is impervious to catastrophe—as history has repeatedly demonstrated. The fact is that there are things—both natural and man-made—that really are capable of bringing about the doomsday scenarios so dear to many end-times prognosticators, so we mustn't make the mistake of deciding that since doomsday prophecies have been wrong up to now that they will *always* be wrong. That would be as presumptuous as looking for our immediate demise, and possibly just as dangerous.

So, just how dangerous is the future? To answer that, it might be a good idea to take a look at some of the genuine dangers we face as a civilization in an effort to determine how much we really need to be worried. In being able to appraise each threat in a realistic and level-headed manner, we can hopefully assuage some of our fears and perhaps, if any of these dangers ever become more than merely theoretical, be able to face them in a more deliberate and panic-free way. The following sampling is not ranked in any particular order of likelihood and probably does not constitute the entirety of threats Earth may one day face, but it should be sufficient to help the reader appreciate both the dangers and the prospects for survival of each of the scenarios.

Natural Doomsday Mechanisms

Destructive mechanisms fall into two categories: natural and man-made—with each containing any number of agents capable of destroying an entire global civilization, or at least sending it back to the Stone Age. In determining an agent's capacity to instigate doomsday, however, it is first necessary to understand exactly that we are talking about a mechanism that has the potential to destroy a global civilization and not just isolated pockets of it. In other words, for doomsday to be realized, the destruction must be global in nature and capable of annihilating not only civilization but all human life.

Fortunately, only a few forces are capable of doing such a thing. To appreciate what these forces might be and how destructive they are, the chart below lists some the more commonly touted agents usually thought to be capable of destroying, in turn, a global civilization, life in general (not just human beings but all life on the planet), and the environment, rated on a scale of 1 to 5 (with 1 representing the lowest potential for destruction and 5 the highest).

GLOBAL DESTRUCTION POTENTIALS

	Potential to Destroy/Significant Impact:		
	Global Civilization	All Life	Environment
NATURAL CAUSES			
Earthquake	1	1	1
Seismic/tidal wave	1	1	1
Supervolcano	1–3*	2	4
Comet/asteroid hit	5	4–5	4–5
Global weather changes	2–3*	2	3–4
MAN-MADE CAUSES			
Conventional war	1–3*	1–2	1–2
Nuclear war	4	3	3–4
Biological attack	5	4–5	1–2
Industrial pollution	1	1–3	3–4

KEY: 1 = Little or no impact 2 = Some impact/light damage 3 = Moderate impact/significant damage
4 = Significant impact/major damage 5 = Major impact/total destruction

* Destructive potential dependent on affected societies' level of advancement and technology.

Looking at the chart, one might notice right away that many of the natural causes traditionally looked upon as evidence of divine retribution generally have little capacity to destroy either civilization or all life on Earth. Earthquakes, for example, while extremely destructive, are localized events that may level poorly built structures and change the flow of a river, but have almost no impact outside of the quake zone itself. This is also true for seismic (commonly but mistakenly called *tidal*) waves, which, while capable of inundating miles of coastline and devastating a coastal city, could hardly destroy all civilization on the planet. Even the largest tsunamis lose their punch shortly after making landfall, and while momentum may push them inland a few miles, they quickly collapse of their own weight and recede back into the sea—though usually not before doing tremendous damage to both the environment and any populations unfortunate enough to be caught in their path. But tsunamis, like earthquakes, are also localized events that would find their destructive potential limited by geography and other oceanographic considerations. As such, not even a remarkable series of "super waves" could destroy a global civilization.

Volcanism

A volcano, another popular candidate for continent killer, is more potentially destructive (especially to the environment) but unlikely to destroy civilization in its entirety, no matter how big an eruption it is. Like earthquakes and tsunamis, even the largest volcanic eruptions are largely localized events that would impact only those cities and population centers within a few dozen miles of the eruption site. Of course, it isn't the blast itself that does all the damage: big eruptions are notorious for spewing millions of tons of material into the upper atmosphere, which in turn has the ability to impact weather patterns hundreds or even thousands of miles from the eruption site. Normally such effects are short-term and not particularly dangerous, though they can be disruptive and damaging in the short term.

However, there are two exceptions that could spell—if not extinction—certainly great hardship for the planet in general. The first of these would be a series of large volcanoes erupting over a comparatively short span of time, whose cumulative effect could seriously threaten crop production worldwide, potentially resulting in worldwide food shortages and creating significant social upheaval. The other exception would be if Earth were to experience something called a "supervolcano,"[1] one large enough to devastate an entire continent were it to erupt as well as being capable of drastically affecting the planet's environment for decades afterward.

And "super eruptions" are not mere myths either. Our planet has experienced such events with some frequency throughout its geological history. Fortunately, however, they are rare (at least as far as human lifespans are concerned): as far as volcanologists can determine, the last really massive super eruption occurred around 26,000 years ago at Lake Taupo in New Zealand, an event that ejected over 1,100 cubic kilometers of pyroclastic material into the air and probably altered the global environment for centuries to come.

However, Taupo was tiny compared to the eruption at Lake Toba, Indonesia, over 75,000 years ago. That eruption, by some estimates, may have expelled as much as 2,800 cubic kilometers of material into the air (compared with a mere twenty-five cubic kilometers from Krakatoa in 1883 and the puny 2.8 cubic kilometers ejected during Mount St. Helens' famous eruption in 1980) and may have killed 60 percent of all human beings on the planet. Only America's Yellowstone Caldera is capable of putting on such a show, having done so twice in the last two million years, the latest occurring just over 640,000 years ago.

But could a supervolcano going off today truly spell the end of humanity? Probably not. While the effects would be significant and, in areas, devastating, our technological capability would probably enable us to offset the worst effects, at least within the more advanced nations on the planet. Developing nations would bear the brunt of

---

1. A "super eruption" is defined as one ranked with a Volcanic Explosivity Index of 7 or 8 (VEI-7 or -8) and that ejects at least 1,000 cubic kilometers of material.

such an event, likely resulting in death by famine of hundreds of millions of human beings (with livestock deaths numbering in the billions), but within a few decades civilization—battered and bruised, perhaps, but still quite intact—should be back on its feet.[2]

Additionally, though evidence suggests the earth has endured periods of intense volcanism during periods in our past, the last such active period appears to have occurred tens of millions of years ago. Moreover, a dramatic upswing in volcanic activity should also provide plenty of notice beforehand—probably providing volcanologists years, if not decades, of advanced warning. While big eruptions such as Krakatoa (and even more environmentally destructive eruptions such as that of Mount Tambora in Indonesia in 1815) are essentially once-a-century events, the type of super eruptions of doomsday lore are once-every-100,000-year events, making the prospect of doomsday coming in the form of massive volcanic blast and a sky blackened with ash a remote one.

Asteroids, Comets, and Meteors

Of course, volcanoes are not nearly as popular a doomsday scenario as is the prospect of a comet or an asteroid hitting the planet (a spectacular doomsday scenario especially popular with Hollywood), which, depending on the composition and size of the celestial object, really could wipe out a global civilization. Just such an asteroid, in fact, is thought to have exterminated the dinosaurs (along with approximately 85 percent of all species on Earth) some sixty-five million years ago, while similar celestial objects may have been responsible for

---

2. Certainly, such an event would create political instability around the world, but because the most hard-hit victims and, as such, the countries most likely to fall into chaos, are the planet's poorest and least technologically advanced, any conflict or instability that resulted would most likely be low-tech and localized in nature. More advanced nations would be better able to endure such a calamity—especially in the long term—though if global powers such as China, Russia, and India were especially hard-hit, the potential for more serious conflict could be significant.

other mass extinctions throughout history. If we were looking for a natural agent to herald in doomsday, this could well be it.

What makes such celestial objects especially scary is that they can theoretically appear at any time with only minimal warning (in some cases, no more than a few weeks) and, if large enough, could truly trigger a genuine doomsday scenario. Additionally, these objects don't need to be especially large to do damage. According to best estimates, the asteroid that wiped out the dinosaurs was only about ten kilometers (six miles) in diameter, but even at such a comparatively small size it produced a crater over 112 miles in diameter and exploded with the energy equivalent of *100 trillion tons* of TNT, or about two million times greater than the most powerful thermonuclear bomb ever tested!

Space is a big place, making the chances of a large one of these errant rocks pelting the planet pretty remote. While estimates vary considerably over how grave a threat asteroids pose to the planet, the chances of one ending life on the planet anytime soon is variously estimated to be anywhere from one in 10,000 to as low as one in 10,000,000. Of course even a small one—say, no larger than the size of a typical house—is capable of leveling a small city and leaving a crater a mile or more in diameter, so the danger can't be entirely ignored.

Fortunately, most meteors are no larger than a grain of sand, and our atmosphere is very good about disposing of such pests long before they make it to the ground. Dozens of baseball-sized meteorites, however, strike the earth each year and larger ones strike every few decades, so the potential of being killed by one of these wayward space rocks—while astronomically low—is not impossible.

Comets, however, are another story. Despite being composed primarily of frozen gas (which makes them considerably less dense than an asteroid or meteor), they are just as potentially destructive and, in many ways, even more unpredictable. For example, when the comet Shoemaker-Levy 9 hit the planet Jupiter in July 1994, the pieces of that comet (it had shattered and was moving in formation) entered the Jovian atmosphere at an average speed of 130,000 miles per hour (60 kilometers per second) and delivered the energy equivalent of 200,000 megatons of TNT.

In fact, there is evidence that a comet—or, at least, a fragment of one—may have exploded over Earth as recently as 1908. On June 17th of that year, a massive explosion occurred about 25,000 feet over the Tunguska River in Russia, leveling an estimated eighty million trees over a 2,150-square-kilometer (830-square-mile) area and producing a 5.0 Richter scale earthquake capable of being detected on seismic equipment as far away as London. Estimates of the energy of the blast range from five megatons to as high as thirty megatons of TNT, with 10–15 megatons the most likely yield (which is still a force about one thousand times as powerful as the atomic bomb dropped on Hiroshima in World War II). Although science is uncertain whether the explosion was caused by the air burst of a large meteoroid or comet fragment, there is general agreement that the object was no more than a few tens of meters across.

However, the chances of the earth being hit by such a beast are remote, at least according to research conducted by Australian National University astronomer Dr. Paul Francis.[3] Using computer simulations and data from an American military telescope, in 2005 his team found that there are seven times fewer comets in our solar system than previously thought. Previous estimates of the number of comets were based on the work of amateur astronomers, who it was assumed were only spotting about 3 percent of the comets passing close to the earth, the rest being missed because they were in the wrong part of the sky or were too faint to be seen. It turns out, however, that these amateur stargazers were doing a better job at spotting the little balls of ice than anyone had imagined, and were successfully identifying closer to 20 percent of the comets flying through our solar system at any given moment. This means that there are far fewer undiscovered comets than previously thought, making the chances of being hit by an errant one even more remote. "I calculate that small comets," Dr. Francis reported, "capable of destroying a city, only hit the earth once every

---

3. According to a press release from the Australian National University, dated September 7, 2005.

forty million years or so. Big continent-busting comets, as shown in the movie *Deep Impact*, are rarer still, only hitting once every 150 million years or so."[4]

Of course, these results only apply to unknown comets coming from beyond the orbit of Pluto. The earth, however, is still at risk of being hit by known short-period comets like Halley's. As such comets are more predictable, however, if it were determined that one of them were on a collision course with Earth, we would probably have several years—if not decades—to do something about it (such as figure out a way to deflect it), as opposed to a long-period comet, which would probably give us only a couple years notice at most.

Other Natural Threats

Once we move past the prospect of celestial objects and volcanoes as potential doomsday devices, we begin to enter into the world of threats more theoretical than historical. These include such things as rogue black holes, gamma-ray emissions, giant solar flares, the reversal of the earth's magnetic poles, global pandemics, and other dangerous scenarios. What makes these threats more theoretical than practical, however, is that they are largely unobservable or are only potential threats that lack a proven track record of devastating the planet as asteroids and supervolcanoes have. Still, let's take a look at each to see just how much of a threat they may truly constitute.

Gamma-ray bursts

Gamma rays are unimaginably powerful bursts of energy that result from the merging of two collapsed stars. They are so powerful, in fact, that if one occurred as far away as 1,000 light years, it would appear about as bright as out own sun and would quickly cook off our atmosphere and destroy the ozone layer. This would result in ultraviolet rays from the sun reaching the surface at nearly full force, causing skin

---

4. This quote appears online at http://info.anu.edu.au/ovc/Media/Media _Releases/2005/September/070905franciscomets (accessed June 16, 2009).

cancer and, more seriously, killing off the tiny photosynthetic plankton in the ocean that provide oxygen to the atmosphere. The problem is that such a double star is invisible to us, meaning that if one was nearby and did burst, we would have little or no warning, making for a true doomsday scenario.

Fortunately, it appears that such bursts are rare, with the few that scientists have observed over the years occurring in distant galaxies. And considering the vast size of just our own humble little Milky Way—roughly 100,000 light years from end to end—even if one went off in our own galaxy, the chances of it occurring in our own neighborhood is unlikely. Not impossible, of course, but so remote as to be essentially a non-threat.

Rogue black holes

Over the past twenty years or so, black holes—those invisible little vortices of intense gravity left over from collapsed stellar corpses—have become increasingly popular as potential planet killers and for good reason: even a small one—say no more than a dozen miles across—could cause all sorts of problems if it were to make its way through our solar system. So great is its gravitational pull that it would likely alter the orbit of some of the planets and, if it got close enough to Earth, draw us into an elliptical orbit that would create all sorts of extreme climate swings. While the likelihood of it actually colliding with Earth itself would be extremely remote, in a more reasonable worst-case scenario it could toss us out of our orbit entirely and even eject us out of our system. Additionally, based on recent observations—but more on theoretical suppositions—it is estimated that there could be as many as ten *million* black holes in the Milky Way galaxy alone, making the prospect of encountering one a little more likely.

How likely? Well, not very, actually. They don't move fast (no faster than a normal star), meaning that if one approached our solar system we would have decades or, if we were technologically advanced enough, even centuries to notice its approach (which would be done by noticing minor variations in the orbits of some of the outer planets as the beast affected their paths). Of course, there wouldn't be much we could do about it at that point beyond possibly evacuating the planet and heading into deep space, making it more of a long-term doomsday scenario than an overnight one.

Giant solar flares

As every schoolchild knows, our sun is constantly shooting off gaseous plumes of white-hot plasma, which we call *solar flares* or, more properly, *coronal mass ejections*. Fortunately, these enormous magnetic outbursts, which bombard Earth with a torrent of high-speed subatomic particles, are largely negated by our planet's atmosphere and magnetic field, so we seldom feel the effects of these plasmic bursts (beyond creating havoc for ham-radio users and increasing the luminosity of the aurora borealis, or Northern Lights). They are of concern mainly to space explorers, who really would have a problem if caught in orbit without suitable shelter when one of these things go off.

However, astronomers have occasionally witnessed other suns in our galaxy produce something called a *super flare* millions of times more powerful than their regular cousins, which, if it occurred on our sun, would turn our planet into a charcoal briquette. Fortunately, there is persuasive evidence that our sun doesn't engage in such foolishness, and that such massive flares seem to be confined mostly to newer stars or to those demonstrating significant gravitational instability.

Yet our sun may be capable of exhibiting milder but still disruptive solar activity potentially capable of raising temperatures on Earth enough to induce massive flooding or, in the case of decreased solar activity, lowering the temperature enough to induce a mini Ice Age (which would be far more destructive to the world's economy than global warming).[5] This is not end-of-the-world kind of stuff, however, but more of a long-term global change problem.

---

5. In fact, there is evidence that decreased solar activity may have contributed to seventeen of the nineteen major cooling periods on Earth in the last 10,000 years.

Reversal of Earth's magnetic field

It seems that every few hundred thousand years Earth's magnetic field dwindles to practically nothing and then gradually reappears with the north and south poles flipped. Now, this flipping of the magnetic poles—the last time it happened was about 780,000 years ago—isn't particularly dangerous, but this brief period of decreased magnetic fields—about a century or so in duration—would threaten life on the planet. Without magnetic protection, particle storms and cosmic rays from the sun, as well as even more energetic subatomic particles from deep space, would strike Earth's atmosphere, eroding the already beleaguered ozone layer and causing all sorts of problems to both man and beast (especially among those creatures that navigate by magnetic reckoning). Moreover, scientists estimate that we are overdue for such an event, and they have also noticed that the strength of our magnetic field has decreased about 5 percent in the past century, possibly signalling that such a shift may be in our immediate future—within a few centuries if not sooner—so this is not an unreasonable concern.

However, because it would be so gradual, should scientists in the future discover that such a shift is in the works, there should be plenty of time to take the necessary precautions to avoid the most destructive effects by moving underground or off planet, or perhaps by strengthening the planet's atmospheric defenses with the use of exotic, futuristic technologies. In any case, it isn't something we need to worry about in the short term—though it could be a concern for those living a few hundred years from now.

Global epidemics

Germs and people have generally managed to coexist in peace for thousands of years, but occasionally the balance gets out of whack and all sorts of unfortunate events result (such as the Black Death, which wiped out a quarter of the population of Europe in the fourteenth century, and the influenza outbreak in 1918–19, which took at least twenty million lives worldwide). As such, the prospect of a new antibiotic-resistant germ developing in nature that could decimate the world's population again is a very real concern not to be taken lightly, especially considering the speed at which diseases can be spread nowadays. The grimmest possibility would be the emergence of a strain that spreads so fast that we are caught off guard or one that resists all antibiotics, perhaps as a result of our stirring the ecological pot. In fact, it is thought quite possible that the sudden wave of mammal extinctions that swept through the Americas about 12,000 years ago may have been the result of an extremely virulent disease that humans helped transport as they migrated into the New World.

However, there is good news as well. Science has come a long way since the fourteenth century—and since 1919, for that matter—in understanding how viruses mutate, as well as in developing effective antibiotics for whatever new strains come along. This trend is likely only to continue into the future as new technologies and strategies for battling the constant assaults from our microscopic predators become more advanced as well, meaning that with luck, we should—at least as a species—be able to stay just one step ahead of any future pandemics. That's not to say that a particularly virulent strain of influenza or an airborne version of the ebola virus might not someday appear to bring death to millions (especially in less developed countries), but the prospect of a single bug wiping out all of humanity overnight remains extremely unlikely—though not, of course, impossible.

Human Doomsday Mechanisms

Of course, nature isn't the only force capable of ushering in Armageddon. Humanity has repeatedly demonstrated that it is quite capable of mass destruction when it sets its mind to it. As evidence, consider that in just the twentieth century alone more people died in wars than have died throughout recorded history in all the earthquakes, volcanoes, and tidal waves *combined*. It seems then that what nature is generally loath to do, man is more than willing and increasingly capable of pulling off.

Certainly, the case can be made that our ingenuity has made us the only species on the planet capable of devising the means of obliterating ourselves, but what, exactly, are these man-made mechanisms of our own demise, and just how destructive are they? Here is just a sampling of the many potential human-created doomsday mechanisms to consider.

Thermonuclear war

Probably the first man-made threat that comes to mind when considering human-caused doomsday is the nightmare scenario of full-scale intercontinental nuclear war. In fact, the possibility of such a global conflagration taking place has been the leading preoccupation of doomsday prophets since Alamogordo, and it remains a shadow we have been living under for over sixty years. Additionally, on a planet bristling with literally thousands of warheads—some of them held in the arsenals of governments of questionable stability—and with other developing countries in a race to produce their own weapons, such a possibility cannot be taken lightly.

But could a full-scale thermonuclear exchange really eradicate all civilization on the planet, or are we overestimating such a nightmare scenario's true destructive potential?

At the risk of sounding overly optimistic (and perhaps even a bit Pollyanna-ish), the best computer models consistently demonstrate that despite the immense damage a full-scale nuclear war would inflict upon humanity and the environment, it is unlikely to wipe out civilization *in its entirety*. It would set it back several decades to be sure, and in some especially hard-hit areas civilization would have to essentially start over from scratch, but most likely humanity would survive, especially in the more remote areas of the globe. Additionally, much of the military and political infrastructure of the devastated countries—being largely mobile or protected in special facilities—would likely survive as well, thereby providing a basis from which to rebuild. While the industrial base would be shattered and the financial foundation of the global economy would be in tatters, as long as the basic knowledge and technological expertise acquired over the centuries survived, civilization would be able to rebuild. Certain areas might be rendered uninhabitable for years by radiation, and the death toll might well be in the hundreds of millions—if not billions—when all is said and done, but with a world population rapidly approaching the seven billion mark, there would probably be far more survivors than victims of even a global Armageddon.

It could also be argued that with the collapse of the old Soviet empire, the worst of the Cold War fears were, if not eliminated, at least dramatically reduced, making the prospect of either an intentional or accidental nuclear exchange occurring today less likely. Of course, there are still tens of thousands of active warheads on the planet, but most of these are in the arsenals of allied nations and democracies—or at least under the control of stable governments. There are a few unfriendly or unstable countries that either possess a handful of such weapons or are currently attempting to acquire them, so there does remain a real danger that such weapons could potentially still be used someday, but considering human nature's powerful survival instinct, it's unlikely that even the most irresponsible government would attempt to use such weapons except under the most extraordinary circumstances, although one never knows.

Nuclear winter

The real danger from a thermonuclear war is not the blast damage and radiation. These would be fairly localized (most nuclear devices have a blast radius of under ten miles), and the radiation levels would drop fairly quickly to livable levels in most areas within months. What would have the greatest long-term impact would be the effect such an exchange would have not on cities and population centers, but on the earth itself. The detonation of hundreds or, potentially, even thousands of nuclear warheads within the span of a few hours would have a profound effect on the atmosphere, and could usher in something scientists refer to as *nuclear winter*.

While a controversial and not completely understood phenomenon, nuclear winter is the theory that the secondary effects of a nuclear exchange—the dust, smoke, soot, and ash thrown into the atmosphere from the nuclear detonations and the fires they ignite—would do the real damage, at least in the long run. In effect, the clouds of smoke from raging fires, the steam from warheads detonating over open water, and the immense amount of dust thrown up by thousands of nearly simultaneous detonations would be carried high into the upper atmosphere by the prevailing winds until they formed a thick blanket around the planet. This layer of dust and soot would be so thick, the hypothesis maintains, that little sunlight would be able to penetrate the gloom, resulting in a dramatic drop in global temperatures along with the sudden inability of plants to convert light energy into chemical energy via photosynthesis. This, in turn, would have a dramatic effect on Earth's intricately balanced ecosystem and agriculture, resulting in the devastation of the world's food supply and initiating a worldwide famine on an unimaginable scale. This famine, then, when combined with the effects of radioactive fallout, acidic rainfall, and drought, would conceivably kill billions more than would have died in the initial nuclear exchange itself and would potentially bring humanity to the very threshold of extinction.

However, the theory has a few problems. First, we simply don't know how much smoke and dust a full-scale nuclear exchange would throw into the atmosphere, how evenly distributed it would be, or if it would really be thick enough to prevent most sunlight from reaching the earth's surface. Though we have evidence that large amounts of dust and other particulates in the upper atmosphere will decrease the amount of sunlight reaching the surface—the explosion of Mount Tambora in Indonesia in 1815 (the largest volcanic eruption in modern history), for instance, threw up so much ash that it ushered in a "mini Ice Age" that devastated crops in New England and Europe for months afterward—that doesn't necessarily mean that a man-made catastrophe like nuclear war would do the same.

It must also be remembered that Tambora ejected more than 160 cubic kilometers of material when it blew. The possibility that even 20,000 warheads exploding at once could put that much material into the air is extremely unlikely (especially as most of these would be airborne detonations unlikely to throw great amounts of soil skyward). Most of what would rise into the atmosphere would be smoke and dust rather than the heavier material ejected by a major volcanic eruption, making the denseness of the "blanket" considerably less than the ash cloud that would result from a super eruption.

It's also likely that such a plume of smoke and dust, driven by generally lateral air currents, would be largely confined to the latitudes in which the majority of the detonations took place, leaving most areas (and, probably, the poles) mostly clear. As such, it's uncertain how temperature over the entire planet would respond when sunlight is still capable of reaching large areas of the surface. Additionally, as the dust and smoke particles are heavier than the surrounding air, the clouds would dissipate fairly quickly—probably within just a few weeks—as the heavier particles fell back to Earth, eventually clearing the skies and allowing a degree of equilibrium to return to the planet. Of course, the clouds of smoke and ash would have profound detrimental effects on the ecology of the planet and drastically alter weather patterns for decades afterward, but that these effects would be capable of extinguishing an advanced, global society in its entirety is uncertain and still open to debate. As most government and military institutions would survive, along with protected and privileged elements of the populace, the seeds of a rebuilding effort would be secure. It may take decades for some degree of normalcy to return, but any technologically advanced civilization should eventually be able to rise from the ashes of its own stupidity to start the rebuilding process over again.

Biological weapons

The price of progress over the years is that today science has the capacity to produce weapons potentially deadlier than even nuclear weapons. These deadly agents aren't capable of producing massive explosions nor are they radioactive in nature. In fact, most are even too small to be seen with the naked eye, but in the wrong hands they really could spell doomsday for humanity under the right conditions. Of course, I'm referring to biological weapons—"superbugs" in generic terms—which really do posses the capacity to kill billions of people, making it a genuine nightmare scenario if they were ever to be used.

It's important for the reader to understand that we're not talking about chemical agents such as tabun, VX, or sarin gas. While these weapons have the capacity to kill thousands of people were they to be unleashed on a major population center, nerve agents are not capable of obliterating a global civilization for two reasons: first, such agents can be difficult and expensive to produce, especially in the massive quantities required to blanket entire cities; and second, they tend to dissipate very quickly (especially in wind or rain), making them only marginally effective. While their use in a confined environment such as a subway, for instance,[6] has the potential to kill hundreds or even thousands of people in a few minutes, their use in an open environment would probably result in more people being killed fleeing in panic than from the gas itself.

However, biological agents are a different story, for such weapons truly do possess the capacity to destroy civilization on a global scale. Like a naturally occurring pandemic, an artificially produced virus—a designer "superbug"—could cut through a population like a scythe, leaving cities and towns intact but utterly devoid of life. If it spread quickly enough and was especially virulent, it could eradicate the entire population of the earth in a matter of months, leaving our planet entirely devoid of human life. Considering the fact that there are terrorists out there willing to give their lives for their "cause," the prospect of such a genetically engineered germ falling into their hands—and of them actually using the thing on innocent civilians—remains a reasonable cause for alarm. It wouldn't be an overnight doomsday, to be sure, but if it ended in the extinction of humanity, it would be, for all practical purposes, the end of time as far as *Homo sapiens* is concerned.

Fortunately, however, biologically exterminating an entire population is not easy. It takes a high degree of scientific and technological sophistication to create a viable agent, along with a high degree of foolhardiness to use it. People smart enough to produce such a germ

---

6. By way of an example, a religious cult in Japan unleashed sarin gas in the Tokyo subway system in 1995, killing twelve people and injuring one thousand, so the threat such agents pose cannot be overestimated.

are also smart enough to know it could come back to haunt them as well, so the reluctance to use such a weapon would be considerable. Unfortunately, such restraints may not stop a terrorist bent on suicide, although even terrorists should still recognize that the extermination of all life would be counterproductive to their goal; after all, one cannot install a theocracy on Earth or usher in a new age if there is no one left around to be subjected to it. Moreover, there would still be the question of how an individual would come to possess such an agent in the first place, as it would still require a sophisticated technology to produce the weapon and a great deal of stupidity to sell or give it to a terrorist organization, so the same prohibitions apply.

Additionally, as is the case with naturally occurring pandemics, science should have the capacity to counter the effects of such a "superbug," were one ever to be unleashed. This idea works from the premise that any civilization capable of producing such a virus would presumably also possess the technology necessary to counter it. In effect, the higher the technology level, the more deadly the germ, while at the same time the smaller the chance the germ would be successful due to an advanced civilization's ability to counter it.

Biotech disaster

One of the remarkable advances science has made recently is in the field of genetic engineering, which, when done correctly, can make crops hardier, tastier, and more nutritious. Engineered microbes can also ease health problems, and gene therapy offers the promise of repairing elusive fundamental defects in our DNA, making the entire field an area of great potential. But there's a danger that genes from modified plants might leak out and find their way into other species, creating all sorts of unanticipated and unfavorable mutations, while engineered crops could foster insecticide resistance, making them especially vulnerable to pestilence. And of course there's always the possibility that a terrorist group might use such a technology to engineer an especially destructive weed (one capable, say, of destroying entire crops) or even produce a mutated form of influenza or some other lethal pathogen. Fortunately, like bio-weapons, such pathogens would take an extremely sophisticated laboratory to produce, and how adversely they might impact our planet remains to be seen, though it seems one would be wise to err on the side of caution when the manipulation of DNA is concerned.

Particle accelerator mishap

Another often overlooked consequence of science is the possibility that one of our high-tech experiments might go awry, with all the unfortunate consequences that would entail. One of the more unusual (and, perhaps, unexpected) concerns is the idea that a particle accelerator—those delightfully strange and super-expensive donuts that fire atoms at each other at light speed in an effort to see what happens when they collide—could set off a chain reaction that would destroy the world by inadvertently creating a subatomic black hole that would slowly nibble away at our planet until eventually it was reduced to a cloud of dust orbiting the sun.

However, just as fears that the first atomic bomb would start a sustained reaction that would set the atmosphere on fire proved to be unfounded, scientists claim that such concerns about particle accelerators are likewise unfounded. The accelerators in use today simply aren't powerful enough to make a black hole, and the fact that accelerators have been in operation for decades without a single black hole to show for them should also be seen as an encouraging sign. Of course, maybe it's just a matter of building a big enough donut....

Nanotechnology mishap

Though it has all the earmarks of a *Star Trek* episode, the fact is that science is just a few decades away from creating self-replicating microscopic nano-machines capable of performing surgery from inside a patient, building any desired product from simple raw materials, and doing other equally spectacular things otherwise beyond our capabilities. Clearly, nanotechnology has the potential to revolutionize manufacturing, medicine, and technology in a way scientists could not have imagined a few decades ago, making it an exciting and promising new field of study and one likely to grow in importance over the coming years.

The problem is in what we don't know about this emergent technology. What if, for example, as a result of an industrial accident, bacteria-sized micro-machines known as *nanites* spread through the air, replicated swiftly, and ended up reducing the biosphere to dust in a matter of days? Plus, as they would also make superb military weapons (imagine what an army of nanobots could do to an enemy radar system or the inner workings of a nuclear submarine), consider how destructive they would be in the hands of terrorists. Clearly, such weapons would be disastrous and practically unstoppable.

Most who work in the field, however, consider the threat overblown. Like the biological warfare scenario discussed earlier, such a weapon—either in the hands of the military or a terrorist cell—would prove too unpredictable to use, offsetting whatever advantages it may bring to the battlefield. Additionally, it shouldn't be difficult to figure out how to program the little devils to turn themselves off after a fashion, precisely to prevent such a destructive scenario from occurring.[7] Still, one can never be certain where such technology will lead us or what it might be capable of doing—nor can we guarantee we'll always retain control over it.

Artificial intelligence

Speaking of nanotechnology, another related technology has to do with the development of smart machines capable of thinking for themselves, à la Lieutenant Commander Data of *Star Trek* fame. Of course, we're not talking about anything quite as sophisticated as Data, but weapons—missile guidance systems, say—that are capable of "thinking" of ways to confound an enemy's defenses aren't far away, nor are smart drone fighters capable of learning from previous engagements and adjusting their tactics accordingly. Clearly, thinking machines would come in very handy in a number of ways, and they would revolutionize our planet if they were to come to fruition.

---

7. It might even be possible to create anti-nanite robots designed to hunt down and destroy the "bad" nanites, thereby solving the problem.

The problem, of course, is what if these smart machines got smart enough to decide that we're no longer necessary and did away with us clearly inferior biological units? With their artificially enhanced reflexes and faster thinking speed, they might prove a formidable enemy that really could prove dangerous if there were enough units involved—an idea demonstrated in graphic detail in the 2004 film *I, Robot*. Of course, that was Hollywood's vision of the future, but the prospect of a world awash in increasingly intelligent machines may be only a few decades away, making such a scenario not entirely beyond the realm of possibility—especially when one considers man's occasional short-sightedness when it comes to the lethality of his own inventions.

But how great a danger is artificial intelligence? It all depends on how intelligent it is and what sort of safeguards might be built into it. First, developing an artificial intelligence sophisticated enough to decide to do away with us "mere humans"—if, indeed, such a technology were even possible—is still a ways off (think centuries, not decades) and so not something we need to worry about for a while. Second, it should be possible to program a fail-safe system into any artificially intelligent machine we create, forbidding it to harm human life (a concept proposed by science fiction author Isaac Asimov as early as 1942); and, finally, even if they were to take over, who's to say that smart machines might not make better leaders than humans? Seriously, though, while such a prospect may be a concern for future generations, it isn't something we need to worry about for the time being—if ever.

However, a bigger concern would be the prospect of humans merging with machines—that is, of humans possessing the ability to essentially download their brains into computer-enhanced mechanical surrogates, which would allow them to log in to nearly boundless files of information and experiences, making it possible for them to vastly expand their intellectual power. This would make such a person potentially capable of far more than even the smartest and strongest human, and making them particularly dangerous villains if they chose to go over to "the dark side." However, such a technology would probably be carefully regulated and monitored, and it's likely that such "postbiologicals" would still need regular human maintenance to keep them going, so this prospect is not a given. Also, it's not clear how a single cyborg—even a remarkably clever one at that—could threaten all humanity. It would still, after all, be limited physically, and presumably still be capable of being destroyed, which makes the prospect of computers ushering in doomsday most unlikely—at least for the time being.

Environmental toxins

And finally there is the impact from our chemicals and waste products. Certainly the death of thousands of people in an industrial pesticide accident in Bhopal, India, in 1984, serves as a poignant example of just how dangerous chemicals can be. It's also uncertain how the pesticides and thousands of other tons of chemicals we dispose of every day might affect us down the road. In high doses, for example, dioxins can disrupt fetal development and impair reproductive function, potentially damaging our ability to procreate and therefore eventually driving humanity to extinction.

Of course, such a possibility is a very long-term doomsday prospect and not one most people normally think of when considering end-of-the-world scenarios, but if humanity is to one day die out, this is the most likely path such a die-off would take. However, because we are growing increasingly aware of the problem that chemicals pose to us and our environment and since we are already beginning to take steps to address these threats (such as shifting to non-polluting renewable energy sources), we should have some cause for optimism. We may never be able to omit pollutants from our civilization entirely, but hopefully the future will be far cleaner and safer than the present (which in turn has already proved to be far cleaner and safer than the past).

Conclusions

I realize I may have left a few potential doomsday scenarios out of this list. For example, I haven't addressed the possibility of the universe collapsing in upon itself due to the machinations of the mysterious stuff known as dark matter, or of basalt volcanism turning entire continents into seas of flame and smoke à la the biblical cities of Sodom and Gomorrah, or even of the possibility that massive deposits of subsea methane might bubble to the surface as the Arctic region becomes warmer, thereby flooding the atmosphere with a gas twenty times heavier than CO_2 and enhancing the greenhouse effect. The point to be made from all this, however, is that while there really are things that could spell the end of us, they are either such rare events or so unlikely to happen in our lifetime as to be, for all intents, nonexistent. If the past is any indication, I suspect we will continue to move on as before, occasionally dodging the stray meteor and enduring gradual changes in the global climate, but reasonably safe and secure on our blue orb in space.

Plus, since we are aware of these potential dangers, it gives us a greater capacity to prevent them from happening, thereby significantly reducing the likelihood of us being victimized by our own technologies. However, as the wise sage Murphy repeatedly reminds us, eventually whatever can go wrong will go wrong, so we need to be constantly on our guard.

Some, however, accuse Murphy of having been an optimist, with which I must agree. I believe that the end *is* coming, and that's not merely an opinion but a scientific fact. Whether we get smashed by an errant comet, blasted to bits by a moon-sized asteroid, succumb to biological or nuclear weapons, or are eradicated by nanites capable of taking over the world remains to be seen, but even if we manage to avoid all of these fates, our sun will one day use up its last reserves of fuel and expand to a size that will incinerate the inner planets (ourselves included), leaving this great blue-green sphere we call home nothing more than an ice-encrusted cinder floating amidst the inky blackness of space.

Of course, we have some time yet to prepare—about four billion years or so—but the fact is that the end is already preordained. And when it does come, I have no doubt that some self-proclaimed prophet or psychic somewhere will take credit for having predicted the end with such uncanny accuracy, giving him or her the last laugh over an unbelieving humanity—or, at least, what's left of it by then.

Chapter Twelve

Ecological Doomsday

While many doomsday scenarios can be dismissed as the ramblings from the religious, New Age, or psychic communities or as highly speculative celestial or geological events unlikely to occur in the near future, there is one type of doomsday scenario that registers with more people than any other, both because of its immediacy and because there is hard science to back up much of it. That is, of course, the popular and growing contention made by many in both the environmental and scientific communities that we human beings are manufacturing our own doomsday by poisoning the very air we breathe, and harming the atmosphere with the pollutants and greenhouse gases we spew into the air each day. In other words, we may literally poison ourselves into extinction!

And it's not just fear-mongering and what-if theories behind all the conjecture; the idea that we might be destroying Earth's delicate ecosystem through our shortsightedness is a very real and growing concern. The destruction of the ozone layer, for example, would have extensive and far-reaching effects on humanity in terms of increased cases of skin cancer, while toxins in our water supply could impact us on a genetic level, threatening the very ability of our species to procreate and dooming it to eventual extinction. The last few decades alone have done much to demonstrate that deforestation, increased CO_2 levels, and unregulated urban sprawl threaten to so dramatically alter our weather patterns that within a century our planet might be a much less hospitable place to live.

Even more urgently, due to rampant and largely wasteful human activity—now compounded by China and India as they rapidly industrialize—greenhouse gases (most notably CO_2) are building up in the atmosphere at a prodigious rate, thereby trapping more heat from the sun's rays beneath an invisible "blanket" of gas. This "greenhouse effect" is conjectured to drive up worldwide temperature averages, resulting in the rapid meltdown of the polar icecaps that in turn will, due to the influx of billions of gallons of melted polar ice, raise world sea levels, potentially submerging low-lying areas of Europe, America, and Southeast Asia. Climatologists speculate that the rise in global temperatures will also render certain areas of the planet, particularly parts of sub-Saharan Africa, too hot to grow crops, thereby reducing total arable acreage and causing widespread famine among the planet's poorest nations. Moreover, increasing competition among industrialized nations over dwindling food and energy sources could trigger regional and even global wars—wars that could conceivably go nuclear as more countries acquire weapons of mass destruction. In a worst-case scenario, hundreds of millions of people could die either from starvation or from intermittent warfare, with possibly billions more being displaced or otherwise adversely affected and driven to ever more desperate acts in an effort to survive.

Of course, the prospect of environmental disaster has been a popular theme since the 1960s, but it wasn't until the start of the new millennium that such concerns took on global implications and interest in global warming and the conjectured political/social chaos resulting from it began to be taken seriously. And although global warming has been championed by a number of high-profile celebrities for years, perhaps no one has done more to bring this scenario into the public consciousness than former U.S. vice president and Nobel Peace Prize winner Al Gore, whose Academy Award-winning documentary *An Inconvenient Truth* has, along with both its companion book and Gore's earlier book *Earth in the Balance*, won over many converts and continues to be an influential force behind the controversy today. In fact,

An Inconvenient Truth has not only won critical acclaim, but has subsequently become almost mandatory viewing in many schools and colleges across the country, demonstrating just how mainstream the belief in our own future demise has become.

And it's not just politicians and celebrities who have embraced this theme, but many mainstream scientists as well, who have been arguing for over a decade that the earth is moving into a period of warming that is likely to have significant impact upon the world—a basic conclusion that has since been endorsed by at least thirty scientific societies and academies of science, including all of the national academies of science of the major industrialized countries. While individual scientists have voiced disagreement with some findings of the IPCC (Intergovernmental Panel on Climate Change),[1] these scientists are clearly in the minority. The overwhelming majority of researchers working on climate change agree with the IPCC's main conclusions, providing the entire theory with an impressive body of credentialed authority.

However, before we surrender to despair over these very real possibilities, let's take a moment to consider the entire issue from a more objective perspective. After all, with the exception of a confirmed rise in world temperatures over the last fifty years and a demonstrable retreat of many glaciers around the globe, many of the consequences being touted by the theory's most ardent proponents have yet to materialize, inviting us to stand back and examine the issue in a more dispassionate manner (which is something that is becoming increasingly difficult to do in the current highly charged political and emotional climate that continues to swirl around the issue).

So just how great a threat *is* global warming? Are rising temperatures and sea levels really going to spell the end of civilization, as some

---

1. The IPCC is a scientific body that was established in 1988 by the World Meteorological Organization (WMO) and the United Nations Environment Program (UNEP), and is tasked with the job of evaluating the risk of climate change posed by human activity.

insist, or is the danger being overblown or, in some cases, even entirely fabricated?

The Great Debate

While initially it appeared that the entire international science community was onboard in predicting a man-made apocalypse just around the corner, as the science became better a degree of caution with regard to some of the more spectacular claims began to emerge. For example, more recent and accurate data appears to demonstrate that while ice sheets in the Arctic and around Greenland are indeed melting faster than they are being replaced, the opposite appears to be the case in Antarctica where, according to a 2002 study conducted by Ian Joughin of NASA's Jet Propulsion Laboratory and Slawek Tulaczyk of the University of California, Santa Cruz, the ice sheets appear to be *growing*, largely offsetting the amount of melting in the Arctic.[2] Additionally, it has been pointed out by climatologists that CO_2 gas—opined by many to be the leading cause of global warming—is not responsible for the rise in global temperatures at all, but is in fact the *result* of the rise in temperatures. In other words, CO_2 levels (which are a natural byproduct of human and animal life and not a pollutant) rise as a *result of the planet growing warmer*, not the other way around.

Moreover, it has been suggested that some of the various scenarios being argued among scientists are based upon flawed, incomplete, or, in some cases even "massaged" data, thus allowing for widely divergent suggested outcomes—outcomes that are more often dictated by a particular political or scientific agenda than by hard science. While we like to imagine science to be apolitical and dedicated purely to the pursuit of the truth, the fact is that science is subject to many of the same foibles and biases that are common to most areas of human endeavor. Certainly, once global-warming research became politicized and a source of income to many organizations, universities, and think

---

2. *New Scientist* magazine, January 2002 (vol. 295, p. 476).

tanks (more than twenty-five billion dollars in government funding has been spent studying the issue since 1990), the pressure to reach politically popular consensus became intense and made any challenges to the "official" findings increasingly problematic and potentially career ending.

Another problem with the global-warming hypothesis is that much of it is based on computer models, which have been repeatedly demonstrated to be only as good as the data used to create them. A recent example of an unreliable computer model was reflected in the predictions by forecasters at Colorado State University and the U.S. National Oceanic and Atmospheric Administration (NOAA) that, according to their best and most accurate models, 2006's hurricane season would be among the most active on record, with as many as seventeen tropical storms and five major hurricanes with winds of at least 111 miles (179 kilometers) per hour being likely to form in the Atlantic basin—predictions possibly fueled by the unusually destructive 2005 season and the still-deep-seated emotional scars produced by Hurricane Katrina's pounding of New Orleans in August of that year. In reality, 2006 turned out to be among the quietest years on record in terms of major hurricanes,[3] forcing many forecasters to significantly scale back their estimates and scramble for explanations as to how their predictions could have been so far off the mark.

Of course, none of this means that global warming isn't occurring, but it does demonstrate that science can only speculate with regard to how serious the problem is, to what degree human activity may be exacerbating the problem, how extensively it is likely to impact the global environment (for example, estimates of how much the average global temperature may increase over the next century range anywhere from 2°C to as high as 6°C, depending on the model—a huge variance, scientifically speaking), and the extent of the impact on the geopolitical situation. Clearly, if the rise in seawater levels is gradual—say, a quarter-inch

---

3. Only nine tropical storms and two hurricanes with peak winds of 120 miles (193 kilometers) per hour or more formed that year, and both stayed well offshore from the U.S. mainland.

per year throughout the rest of the century—the immediate impact on society will be negligible and probably easily countered by the construction of more robust seawalls and other coastal reclamation technologies. If it proves to be more rapid, however—say an inch per year for the next ninety years—then we could be looking at significant population displacement and loss of real estate.

Additionally, environmentalists often overlook one element of the world-warming scenario, which is that the rise in global temperatures could have beneficial effects. For example, even a small rise in worldwide temperatures would force a retreat in the permafrost covering much of the Arctic regions of Canada and Siberia, thereby creating more arable land in areas traditionally too cold for farming and increasing agriculturally useful acreage worldwide. Climate alarmists often overlook the fact that one of the most agriculturally prosperous periods in European history occurred during the Middle Ages (unusually warm temperatures befell Europe from the tenth to the fourteenth centuries), when milder temperatures and greater rainfall allowed for a larger range of crops to be grown much farther north than had previously been the case.

This unusually warm period even left much of the coastline of Greenland free of ice, making it possible for the Vikings to establish colonies on the island, where they flourished for centuries. It was the advent of global cooling in the sixteenth century—frequently referred to as the Little Ice Age by climatologists today—that ushered in a period of frequent famine and crop failure, demonstrating that it is a *downturn* in temperatures, not an *upturn*, that proves to be the more disruptive and politically destabilizing event. In fact, it could be argued that the advent of advanced civilization on the planet wouldn't have even been possible until the world went into a warming cycle around 18,000 years ago, initiating the retreat of the mile-thick sheets of ice from Europe and North America and ending the last great ice age that had held the world in its grip for tens of thousands of years. Whether the present cycle of global warming is a positive or a negative is largely a matter of conjecture. It is politics that appears to be largely driving the debate today rather than science, making any further discussion of the subject more an exercise in political dialogue than scientific debate.

Further muddying the waters is the recent assertion by some climatologists that due to a marked decrease in sunspot activity and other factors, the world may actually be headed for a period of global cooling—in effect, instead of worrying about the world growing too hot, we may instead need to prepare for it growing too cold—a prospect that could well prove to be in many ways even more disastrous than global warming. Some scientists have even gone on record as saying they believe there's a 50/50 chance that the planet will either turn into an oven or go into the deep freeze in the next few decades. Now that's what I call prognosticating!

Ultimately, the problem with pollution as a doomsday scenario is twofold. First, it is a very slow process that can take decades before its detrimental effects can be fully appreciated and, second, it fails to take into account humanity's ability to successfully respond to such an event. Clearly, it is inconceivable that any sufficiently advanced civilization would knowingly and willingly permit itself to be destroyed by its own toxic waste. As a civilization evolves and becomes more technologically sophisticated, not only would it possess the means to detect the detrimental impact industrial pollutants were having, but since human beings are driven by a strong will to survive, humanity would inevitably find the political will to take the appropriate actions when the evidence of mounting danger becomes obvious and indisputable.

Evidence for this can be found in the Montreal Protocol of 1987, which banned fluorocarbons from industrial use after it was demonstrated that CFCs (the active ingredient in fluorocarbons) were harming the planet's protective ozone layer, thereby increasing the risk of skin cancer to humans. The subsequent "healing" of the ozone in the succeeding decades has demonstrated that nations can, when necessary, come together in a quick and efficient manner to solve a global environmental problem. In this example of widespread cooperation, U.S. private industry, the federal government, the military, and environmental groups teamed up to solve the problem, making a seamless transition from ozone-depleting chemicals to ozone-friendly ones in just a few years.

Furthermore, there is already a global effort underway designed to regulate everything from the destruction of the rain forests to waste management, while the effort to introduce cleaner energy sources and redouble recycling efforts continues to build momentum. As such, it seems likely that the necessary changes designed to offset the worst of the potential problems created by global warming are already underway, and will only become more robust in the next few decades as the extent of the dangers become better understood.

Although financial and political pressures might retard progress in some areas for a time, eventually human beings will demand that the necessary changes be made, thereby overriding natural bureaucratic hesitancy to take quick and decisive action. It is unrealistic to imagine that humanity will stand by apathetically and watch itself commit ecological suicide (though significant damage could be wrought before serious action is taken); the instinct for self-preservation is too strong and will ultimately override the instinct for personal enrichment. I could be wrong about that, of course, but I would be surprised if we destroyed ourselves as a species merely for the opportunity to make a quick buck.

Conclusions

In the end, the fact is that science simply doesn't know what the "proper" temperature for the planet should be, nor is it capable of providing any sort of useful timeline for when we might expect these drastic changes in weather patterns to occur. Moreover, science can't be sure whether the global-warming cycle is part of a naturally occurring cycle of the planet—as seems likely—and how much of it, if any, is caused by human pollution. And, finally, it fails to take into account humanity's resiliency, ingenuity, and flexibility in adapting to changing global weather conditions and creating solutions to counter them.

Obviously, it is essential that we be good stewards of the planet entrusted to us and future generations, and to make every effort to use our natural resources wisely. It is also wise—and for the first time in history, economically and technologically feasible—to shift from a petroleum-based economy toward cleaner and inexhaustible renewable energy sources. However, the current atmosphere of fear and doom being generated by so many scientists and politicians can only be considered an additional burden rather than a real answer to the very difficult challenges created by our planet's constantly changing climate. In the next few years, cooler heads will hopefully prevail and science will be able to separate the hype from the truth, so that appropriate actions can be taken in a sane, reasonable, and timely manner. Otherwise, we will be in danger of frittering away valuable resources and time in an effort to find solutions for problems that do not actually exist anywhere outside of some doomsday environmentalist's computer model.

Chapter Thirteen

The Extraterrestrial Hypothesis

Not all end-of-days scenarios see our planet going out in a "blaze of glory." There are those who anticipate a more positive outlook for humanity—perhaps even a golden age on the horizon, if we can just make our way through the challenges of the next few decades (or, by some accounts, centuries). While many see this as a spiritual awakening among humans, there are those who believe this utopian world will be brought to us by "external" forces. In essence, some anticipate that the end of the present era will be heralded not by angels blowing trumpets or through some sort of massive Harmonic Convergence, but by the arrival of those from beyond the stars.

While this idea is usually met with derision by most people, as science continues to expand the horizons of our universe and humanity is coming to appreciate just how big a place it is, we are becoming increasingly open to the prospect that we are far from the only advanced civilization in the cosmos (however one cares to define the term *advanced*). As such, the prospect that we have been and may even now be a source of interest to other, more technologically advanced civilizations isn't as laughable as it once was. Additionally, the idea that such a civilization may have a vested interest in preventing us from destroying ourselves—much as we might want to save an endangered animal species from being hunted to extinction—is not an unreasonable one.

However, the prospect of extraterrestrials choosing to intervene in our natural development does bring up a couple of questions. The biggest question is why they would do it—in other words, what would be their primary motivation beyond merely giving in to their more altruistic instincts—and the second question is how do they intervene without creating more problems than their arrival would solve?

Let's look first at the *why*.

Alien Interventionists?

There appear to be two primary reasons why extraterrestrials would want to intervene in our planet's development (assuming they are doing so for our benefit and not coming as conquerors): either they consider us a potential threat and are intent on stopping us before we become a danger to them (assuming we aren't already), or they are here as enlightened saviors, ready to save us from ourselves. Before choosing sides, however, let's examine both hypotheses to see how well they stand up to logic.

The idea that aliens might see us as a threat is an old one. Basically, it works along the lines that with the advent of the nuclear age, we have suddenly become a force to be reckoned with, making it necessary that they "tip their hand" so to speak and force us to behave ourselves—either by persuasion or, if necessary, by force. Probably no science fiction movie captured this idea better than the 1951 classic *The Day the Earth Stood Still,* in which an alien named Klaatu comes to Earth to warn us that the galaxy will not put up with our shenanigans much longer, echoing a theme that has remained popular ever since and has only grown as we have become increasingly technologically sophisticated (and hence more dangerous).

A moment's consideration quickly demonstrates such an idea to be nonsense, however, once we consider the high level of technology any civilization capable of traveling the immense distances between the stars would already possess. Without a doubt, to accomplish interstellar travel would require a technology centuries ahead of our own, which would likely render any weapons we might possess today largely harmless to them. In effect, we should be no more dangerous to an extraterrestrial civilization than a Stone Age tribe armed with blow guns would be to a modern army battalion. They may have concerns about our nuclear arsenals (much as soldiers may be concerned about a booming poison-tipped arrow trade going on among the natives), but in terms of us posing a direct threat to their planet hundreds or even thousands of light years away, I suspect they have little to worry about.

Of course, that's not to say we couldn't pose a legitimate threat in the future, especially once we acquire the same interstellar capabilities they already possess and learn to harness many of the same energy sources they do. However, in the time it takes us to develop a full-blown interstellar capability—say two hundred years—any alien civilization studying us today will also have continued to advance as well, keeping us well behind them in terms of technological equivalency. In effect, we would always be playing a game of catch-up, with our competition consistently remaining several generations ahead of us, thereby perpetually rendering us insignificant as a threat. They may have concerns about what we might do to each other with our weapons, but as far as their own personal safety goes, I doubt they have much to fear from us.

The other hypothesis—the aliens as saviors concept—is the more popular theory and one that I admit to finding far more likely than the idea that aliens fear our technology. It is especially attractive when one considers how much more spiritually evolved an extraterrestrial civilization would likely be when compared to our own and what they could offer us in terms of knowledge and wisdom. In fact, I find such a prospect an exciting one to contemplate and would look forward to such a day if it were to arrive.

However, the savior scenario also suffers from several problems—the most serious being how their arrival would be greeted by the majority of humans. New Agers generally reject the notion that the public would panic if an extraterrestrial civilization made itself known to us, believing instead that the populace has become so used to the idea of aliens that their proven existence would come as little surprise and as such, would likely be readily accepted. I think such a presumption fails to take into account human nature. While no doubt some people would accept the arrival of extraterrestrials calmly—and perhaps even joyfully—many others would not. Consider that there are hundreds of millions or even billions of people around the world who have never given the prospect of extraterrestrials serious thought (or do not believe in them at all); what might their reaction be to such news? Also, what of apocalyptic religious sects that would interpret the appearance of such beings in purely religious contexts? Might not such a revelation ignite the very sort of "doomsday" response within their ranks that other apocalyptic musings have ignited in the past? (Think of Jonestown on a worldwide scale.) And what would be the response of the world's military to the news that our skies were filled with extraterrestrials of unknown capability and intentions? Even if only a tiny fraction of the world's population—and their governments—reacted in fear and panic, that would still constitute tens of millions of people worldwide and place a huge strain on the world's police and military forces, as well as pose a major threat to the stability of many governments.

Moreover, consider the impact the extraterrestrials' arrival would have on religion on this planet in general, especially if it were learned that these same extraterrestrials had been responsible for a number of "incidents" from the distant past that have since been codified into our various sacred documents. For example, imagine what the response would be if it were determined that the Israelites had been led out of Egypt not by God but by extraterrestrials commanding them from the confines of a massive spaceship (hidden within a great cloud by day and a pillar of fire by night), or if Jesus turned out to have been a philosopher/alien intent on bringing enlightenment, or if Mohammad received the Koran not from the angel Gabriel but from an ET intent on bringing order to a land in chaos? The level of cognitive dissonance on the planet would be profound and likely politically destabilizing, with entire faith structures and institutions being shaken to their very foundations. How this might affect some governments—especially those heavily influenced by their religious institutions—can only be guessed at, but the fallout would be considerable.

But even if the ancients hadn't interacted with extraterrestrials, there would still be other profound repercussions to consider. For example, what if after the initial shock had worn off, humanity's next act was to clamor for our newfound friends from the stars to "save" us from aging and disease by insisting that they share their advanced medical technologies with us, or that they solve all the problems we have created through our own short-sightedness and selfishness but are loath to tackle ourselves? Furthermore, might we not hold them responsible for all the plagues and atrocities committed throughout history once we learn they were present when they occurred and possessed the means to stop them but took no action? Clearly, the first alien civilization that decides to reveal itself to us is going to be in for a rough time.

I suppose it's possible such a civilization may consider that the potential benefits of showing up outweigh the possible consequences, but if that's the case, then wouldn't we be in our rights to ask why, if their intention is to save us from ourselves, they haven't done so by now? Are they waiting for things get worse and, if so, why? If they are interstellar missionaries, it would seem they need to make their message known *now*, not decades in the future when there may be nothing left to save. Even Christian missionaries sent to save the "heathens" from eternal judgment didn't travel to the farthest reaches of Africa and then wait years before starting the work of proselytizing; they were usually building churches and converting natives the day they arrived. Would we expect anything different from an alien civilization of saviors?

Obviously the work of planetary savior is far more problematic than many appreciate, and it would be a most dubious rationale for extraterrestrials to cross the many trillions of miles of space just to bring us their gospel of peace and enlightenment.[1]

---

1. Additionally, there's the question of how other alien civilizations might respond to our helpful neighbor's good intentions. Is it possible that guardian extraterrestrials aren't just protecting us from our more malevolent neighbors, but from our more benevolent ones as well?

Alien Invaders?

However, there are those who suggest that an encounter between us and our extraterrestrial neighbors might not be the positive experience many imagine. In fact, some doomsday prophets go so far as to incorporate an armed invasion of Earth into their end-times scenarios, a theme echoed by Hollywood science fiction movies during the 1950s and 1960s, and in even more recent films such as *Independence Day*, *Signs*, and the 2005 remake of *War of the Worlds*.[2] Apparently, and despite forty years of *Star Trek* reruns (which often portray aliens as benevolent in nature), humanity continues to harbor more than a small degree of suspicion, fear, and even overt hostility toward extraterrestrials, as is consistent with its general distrust of anything it considers alien, mysterious, or foreign in general.

Of course, this fear isn't entirely groundless, especially when one considers the highly advanced technology a space-faring civilization would likely have at its disposal. Considering the destruction of which we are capable at our current "primitive" level of development, it is sobering to consider what a civilization even just a couple of centuries ahead of us technologically might be capable of doing.

Fortunately, however, UFOs do not appear to be inclined to turn their "blasters" on us, at least from all outward appearances. If extraterrestrials are here at all, it is logical to imagine they have been here for some time (perhaps many centuries, if some ancient texts are to be taken seriously) and, as such, would have had ample opportunity to dispose of us at any point were conquest their intent. In fact, were they determined to scour the planet clean of our presence—or, at very least, enslave us—it makes no sense that they wouldn't have done so long before we developed the technology (i.e., nuclear weapons, nerve agents, lasers, missiles) that provide us the ability to offer at least token resistance. One doesn't wait for their enemy to grow stronger

---

2. Of course, Hollywood has also produced movies introducing more benign aliens, such as *E.T.: The Extra-Terrestrial* and *Close Encounters of the Third Kind*, but such movies remain in the minority.

before attacking but strikes when they are at their most vulnerable. As such, the main reason we can discount the evil alien scenario is because we're still here.

I suppose it's possible that more malevolent alien species simply haven't arrived yet, so we can't entirely discount the possibility of one day having to do battle with visitors from the stars, but this still seems unlikely to me for one major reason: for an advanced civilization to reach the levels of technology necessary to achieve interstellar flight, it would need to be emotionally, psychologically, and spiritually evolved enough to handle that technology (i.e., wise enough not to turn it on themselves or their neighbors). In other words, while I suppose it's possible there could be extraterrestrial civilizations out there that simply enjoy destroying things (or perhaps consider humans fair game for sporting events as they did in the film *Predator*), I believe such an aggressive race would most likely destroy itself—probably in a quest for power, territory, or wealth—long before turning its vast arsenal upon us. Of course, it would be equally presumptuous to insist that there are no warlike or aggressive extraterrestrials at all, but I suspect that if there are, they are either still at a comparatively primitive level of development—and therefore not yet capable of the level of technology required to make them spacefaring—or they are possibly being restrained in some way by other, more spiritually developed and technologically superior civilizations.

This last point cannot be underestimated. I am convinced that there may well be "guardian" races out there that could be protecting us from less enlightened races, much as the police protect society from criminals as much as possible. In fact, I imagine, if we have been a source of study by other alien races for millennia, it's likely we've acquired a sort of "protected species" status by now, keeping us safe for future study. It's certainly a policy we might adopt were we to one day find ourselves overseeing primitive habitats around distant stars, so it is not unreasonable to imagine other races doing the same.

Conclusions

While the prospect that extraterrestrials are going to make themselves known to us—either as invaders bent on conquering us or as saviors intent on protecting us from ourselves—is an intriguing one, it strikes me as being much more of an escapist mechanism than anything approaching logic. While it is likely there are advanced extraterrestrial civilizations in our universe and that some of them may well be observing us right now, I suspect they are far too wise to intervene in our day-to-day affairs. They simply couldn't be that stupid.

I suspect what makes the alien scenario so attractive is that alien intervention is the lazy man's way out of confronting the challenges and overcoming the obstacles the twenty-first century will send our way. What so many don't realize, however, is that these challenges are what make us grow as a species and evolve us spiritually to new heights of human awareness and understanding, without which we would be doomed forever to remain playing in the sandbox of civilization. We are destined as a species either toward greatness or toward disaster; which we choose will remain up to us—just as it always has. To expect our brothers and sisters from the stars to rescue us only retards the process and forces us to remain spiritual infants.

Chapter Fourteen
The Problem with Foretelling the Future

It's been my experience that when most people talk favorably about the prospect of predicting the future, they rarely stop to ask whether it's even *theoretically possible* to glimpse a series of events that will not unfold for months, years, decades, or even centuries. But I think it's an important question to ask, for if the future can be known to some people, we would be foolish not to listen to what they have to say about it. Even more to the point, by refusing even to consider the possibility that the future might be known to a few gifted individuals, we could be putting our very lives in jeopardy—as well as possibly forfeiting the hopes of future generations. As such, it's imperative that we examine their claims carefully just on the off-chance that a few of them may be genuine glimpses of a future, so we might take action today to circumvent whatever negatives that future may hold.

Of course, predicting the future is not the same as making educated guesses about the future. For planning purposes, we often make predictions about what the next years or even decades may hold for humanity—some of which are often detailed enough that decisive action can be taken based on them. Obviously there is an advantage to predicting future social, political, or financial trends so that we might anticipate and avoid various pitfalls they may suggest. In that respect, it's no different from attempting to predict next week's weather by studying the latest satellite imagery and comparing current conditions to past trends.

With this important distinction in place, then, we must ask ourselves if it is even *theoretically* possible that the future might be known in such precise detail based upon nothing more than intuition, dreams, divine revelation, numeric codes, or the positions of the stars and planets. And, even more to the point, is it possible there may be those rare men and women who have been given the gift of determining those events decades and sometimes even centuries in the future for our mutual benefit or warning? After all, if discerning future events remains outside the realm of possibility, then it doesn't really matter what Nostradamus, Edgar Cayce, the Bible, or a host of other oracles tell us to expect; all of it is wrong, or at best coincidental—a point that is all too often overlooked when considering the credibility of most doomsday prophecies.

So, is it possible to really predict future events? Certainly, science is aware that time is a curious thing that moves at different speeds based upon factors such as how fast we are moving in relationship to each other, how much gravity we are enduring, and a host of other mysteries even physicists are still trying to unravel. Even great luminaries like Albert Einstein recognized the fluidity of this thing called time, opening the door for much speculation about whether it might not be possible after all to gain a peek into the distant future under the proper conditions. Since there is still much we don't understand about our reality and the inner workings of, say, parallel universes and multiple dimensions, it would be presumptuous to dismiss out of hand the idea that future events might not be perceivable to us. The fact of the matter is, we simply don't know.

However, if seeing into the future is possible, there is one major paradox doing so poses that makes the whole idea of soothsaying problematic. This is the idea that time is static rather than fluid—that is, that the future is already set and not something yet to be determined. In essence, in order to predict the future it is necessary that the individual events that make up that future already exist in a timeline and, furthermore, that this timeline be impervious to anything we might do to change those events from our vantage point in the present.

By way of an example, let's say we have the means of discovering that a huge asteroid is going to hit our planet on July 11, 2077—an event that will not only exterminate humanity as a species but leave the planet barely habitable by even the most simple surviving life-forms for tens of thousands of years. And, further, let's say we discover this in 2017, a full sixty years beforehand. What might we do about it?

Probably the first instinct would be to try and devise a means of altering the asteroid's trajectory that, by 2077, may well be within our capabilities. However, the moment we take action we immediately encounter a paradox: if we successfully alter the asteroid's path, we automatically negate the event from happening. In essence, by changing the asteroid's trajectory, we remove it from the timeline; however, the moment we do that, we prevent it from becoming a danger in the first place for, in effect, it *is never going to happen!* But if that's the case, then what rationale do we have to take action in the first place? After all, one does not take action to prevent that which is not going to happen!

This may sound complicated, but it is no different were we to alter a past event, thereby changing our current timeline. Alter any significant event in the past and everything that happens afterward changes our current timeline and, with it, our memories of the "old" timeline. In other words, we wouldn't be aware that we had changed the past because we'd have no memory of what it is "supposed" to be. Changing the future would do exactly the same thing: change the asteroid's trajectory and it never hits the planet, and since it doesn't hit the planet, we would be unable to foresee the event because *it will never happen*. See the problem?

Additionally, even if we could change the future, doing so might result in a far greater disaster than if we'd simply left it alone. For example, let's suppose some seer in 1910 foresaw the rise of the Nazi party in Germany in the aftermath of a great European war that was soon to break out and, with its rise, a subsequent and even more destructive second world war that would result from the Nazis' ascension to power. And, moreover, suppose this gifted sage wrote that the man most responsible for this future carnage would be an Austrian named Adolf Hitler. Now, suppose events transpire as prophesied: the bloody conflict of 1914–18, Germany's defeat, the rise of Nazism in Germany, Hitler's rise to power. Would the world really stand back and let it all unfold as predicted? I suspect that at some point, once it became apparent the seer's predictions were not only accurate but precise enough to take action, it would occur to someone to circumvent these events—probably by killing Hitler before he could take control of the nation.

But what would it mean if the assassination of Hitler was successful? Does World War II never take place? Not necessarily. What if another man takes Hitler's place and the Nazis still seize power, and further, what if this individual proves to be more patient and careful than Hitler? For example, perhaps instead of initiating a world war in 1939 he waits until Germany's armed forces are too powerful to stop, and he not only starts the war at a later date, but because of the additional time he bought to make the Nazi war machine second to none, Germany easily defeats the western allies and even seizes all of Russia in the process. By the 1950s it's not the Soviet Union the United States battles in the Cold War but an even more technologically advanced Nazi Germany, which is now a truly global superpower. In light of such events, the question could well be asked if the world was better off with Hitler dead or alive. In the end, then, the alteration of the original timeline results in an even greater conflict, one that costs the world even more in terms of lives and destruction than what would have resulted had the original timeline remained unaltered.

See the problem with prophecy? If it is precise enough to be useful, it can usually be circumvented, thereby initiating an entirely new chain of events that may prove even more disastrous than the original timeline. Yet if one is able to perceive the new "replacement" timeline—this one perhaps even more disastrous than the one it was designed to prevent—the prospect of circumventing it then becomes possible, potentially resulting in an even more catastrophic future. In the end, it becomes a game that can't be won, no matter how we play it.

Flexible Prophecy

Some so-called prophets get around this difficulty by claiming that the future timeline is not fixed at all but is fluid and therefore subject to change. In other words, when a prophet predicts a future event, it is claimed that such an event is only *potentially* possible and not assured and, as such, changeable, making the predictions more akin to warnings than genuine foresight.

Such excuses have been around from the moment the first prophecies passed without being realized and are still used today with great effect. In fact, it is often the only course of retreat left for seers whose predictions have proved to be woefully inaccurate; in claiming their predictions are purely provisional they are able to make any number of guesses without fear of being held accountable when they fail. It also saves them from being labelled as false prophets, for they can always claim that the problem was not in their prognosticating but in the fickleness of fate and the changeability of the future.

Of course, all of this is patent nonsense. If a prophecy is only one of several possibilities the future holds, for all practical purposes it is worthless. The value of predicting the future comes only in it being definite; if it is only one of several possible timelines, it is no better than a guess, rendering the entire issue moot. In fact, such nonsense is not only logically indefensible but could be dangerous, especially if it induces believers to take certain actions in response to what is nothing more than idle speculation. People have put careers on hold, lost their homes and livelihoods, and even put off marriage and having children in the belief that doomsday was at hand. The damage can be considerable and, in some cases, even deadly (a consequence we examined in chapter 2).

Using Prophecy as Confirmation

Another school of thought about prophecy is that it's designed neither to predict the future nor prevent certain events from happening, but is instead a device useful only as a mechanism for demonstrating that a prophet is truly "sent by God." Some so-called prophets word their predictions in such a way that their predictions only become apparent in hindsight. In other words, a prediction may be left intentionally ambiguous at the time it is made, specifically so it might only be understood in retrospect, thereby confirming the prophet's reliability as a seer and bringing greater weight to all of his or her other teachings. This is, in fact, one of the biggest selling points behind certain religious texts: the belief that if some of the prophecies made in them are subsequently realized, it supposedly demonstrates the entire text to be divinely inspired and therefore worthy of worship.

In effect, then, if Jesus of Nazareth was the realization of several Old Testament prophecies made centuries earlier—as many Christians maintain—then that validates the entire New Testament message (and the Old Testament for that matter), making Christianity the one true religion by default. This is also the approach many of Nostradamus' proponents make with regard to their hero's mysterious quatrains: their supposed historical realization demonstrates the man to have been a true oracle.

Unfortunately, such a precept is as morally indefensible as the "one of many potential futures" explanation. If God foresees a horrible event in our future and we have the means to circumvent it through human ingenuity or desire, it would be cruel not to reveal it to us. By way of an example, if a prophet prophesied that a great plague would ravage the people, but he left out the specifics (where, when, how its effects might be minimized, and so on), would that person truly be demonstrating that he was sent from God, especially if he was willing to permit millions to die to make his point? I don't believe so. In fact, if anything, such a person would be demonstrating his cruelty and disregard for human life, all in an effort to solidify his own position as a seer. Using prophecy as a means to demonstrate one's legitimacy as a seer is a dangerous game to play.

Of course, whether Nostradamus or the Bible—or any of the other countless sacred texts or prophets for that matter—actually do predict the future is a cause for debate. Certainly many prophetic utterances can be twisted to take on various interpretations, and it's not always clear—especially in the case of the Bible—whether particular texts are predictive or postdictive in nature, or even whether the passage is being interpreted correctly at all. The fact is that many predictions that are supposed to have been realized by historical events of the past are open to debate. The case for prophecy being a tool either for determining the future or establishing one's prophetic credentials hangs by a very thin thread, one that often snaps upon closer examination.

Conclusions

In the end, whether the future can be seen or not, we still need to deal with the fact that we are writing that future each day. The reality is that we are all collectively and individually creating a linear timeline with each step we take. Every decision we make (both collectively as a nation or species, and individually as a person) affects that timeline in some small way—not by altering it, but by redefining its direction and momentum. Every time a decision is made or a new direction is determined, it immediately closes an infinite number of potential paths that were open to us but are now closed. This, in turn, provides a new slate of paths to go down and with them an infinite number of new possibilities, each of which will be realized or eliminated with every passing second. And that is what makes the journey we call life so exciting. Were we to see what lies ahead, it would render the entire process pointless and deny us our role in building the future; and in doing that, I believe we would lose a major part of ourselves. As long as we inhabit these physical bodies, we are creatures of time and, as such, citizens of the future. Take that away, and we become mere observers of creation, not participants in it, and that would be the greatest tragedy of all.

Chapter Fifteen

Extinction or Utopia: What the Future May *Really* Hold

One element of end-times beliefs that is frequently overlooked is that some prophets—I admit they are in the minority—see the end of the age not as a time of death and destruction, but as the start of a new and glorious age of peace that will finally allow humanity to put its violent past behind it and move on to ever greater heights of enlightenment. In fact, with the dawning of the Age of Aquarius in the late 1960s and continuing on through the Harmonic Convergence of 1987, some have even maintained that this shift in human consciousness has already begun and is only going to pick up momentum over the next decades—even if that momentum may not always be readily apparent.

While such a belief may seem quaint and even a bit naïve, I do not join the chorus of people who casually dismiss such a notion as New Age ramblings out of place on our contentious planet. In fact, I suspect that the willingness of some to write off humanity so quickly is itself premature and may even demonstrate its own kind of naïveté. Just because so much of our past has been filled with darkness does not mandate that our future must be just more of the same. Humanity has a proven ability to learn from its past mistakes, and so it is not out of the question that we have the capacity to create an even brighter future than many imagine possible. As we discussed in the previous chapter, the future has yet to be written; who's to say that the future couldn't be a better place than the past has been?

Of course, one must always be careful when walking the fine line between the giddy optimism of the New Agers and the cynical pessimism of the doomsday prophets when discussing the future, for both sides have evidence to support their positions. On the one hand, we *are* constantly bombarded with bad news—economic downturns, wars and rumors of wars, political strife, the twin threats of terrorism and environmental poisoning, and the social upheaval that accompanies them—which can make it difficult to see how we might make it through the next ten, twenty, fifty, or even hundred years intact. On the other hand, there is abundant evidence that humanity may be less foolhardy and more resilient than many imagine and, in fact, may even be poised on the precipice of a new and glorious age if we can only see past our own selfishness and fear.

So where does the truth lie? I suspect somewhere between these two extremes. Clearly, humanity has many challenges and difficulties facing it as it moves into the twenty-first century, but then again, has there ever been a time when it didn't? Our generation may have to deal with nuclear-armed terrorists and a dying environment, but our grandparents had to deal with Nazism, the Holocaust, and the Cold War. We fight for equality, universal education, and health care; they fought for women's rights, desegregation, and democracy. While we see *potential* Armageddons everywhere, they actually fought *real* Armageddons with real bullets that spilled real blood. The times and the issues facing each generation may be different, but they are each as daunting as those their predecessors faced.

In my heart, I'm an optimist. I really do imagine that humanity is waking to its own potential and beginning to come alive to the possibilities within it. Where I part company with many New Age purveyors, however, is in the notion that such an awakening could ever have a firmly established start date or be triggered by some as yet to be realized catastrophe, as though enlightenment were a switch one pulls that brings the houselights up to ever increasing levels of luminosity.

Instead, I believe humanity has been slowly but undeniably embarked on this journey for some time now.

Long before the Harmonic Convergence or the Age of Aquarius came along, a core element of civilization has been bringing ever greater levels of light to a darkened world. This movement had no start date, and for hundreds of centuries it appeared as little more than a tiny candle flickering in an ocean of darkness, but it has always been there. Humanity may still be in its infancy, spiritually speaking, making progress difficult to perceive, but progress has been made nonetheless. Like a slow but steady infusion of clean water into a filthy pool, the cleansing effect may not be obvious for a very long time, but one day the last of the impurities will be fully flushed out, and humanity will at last be able to look upon its reflection in the water with pride.

But how can I make such a claim? Isn't it obvious that the world is actually growing worse? With overpopulation, pollution, and the perceived decline in traditional morals and religious belief, many argue that we are actually going *backward* spiritually, rendering as nonsense the idea that humanity is advancing.

Of course, successfully gauging the spiritual state of a particular culture is a highly subjective process, one driven more by personal bias than anything approaching fact, so determining whether our culture is more or less spiritually advanced when compared to the past is a difficult judgment to make. Additionally, we also look at the past through blinders that usually permit us only to see the last few decades, thereby severely limiting the context by which we might perceive history in its entirety. In other words, we are usually so fixated on the recent past that we remain largely ignorant of what things were *really* like in the "good old days."

However, a look at the broad scope of history—encompassing the *entire* span of human existence—can be enlightening and, in many ways, give us reason for hope. So in an effort to end this book on a positive note, let's take a moment to consider just what the future *may* hold if we will only retain faith that humanity can weather the many storms it must endure on its way to spiritual enlightenment.

Poverty and Literacy

Multitudes of people on this planet live out their lives in grinding poverty—a condition that remains pervasive. Much of the world survives on a few dollars per day, and in some countries a few hundred dollars per year. Poverty has been a part of the human condition since *Homo sapiens* first emerged, and it remains with us like an ache that refuses to go away no matter how badly we wish it would.

However, once we accept the fact that poverty, deprivation, and want have always been a part of the human equation, the next question we need to ask is not whether there is still hunger and poverty, but whether the *percentage* of people who live in these conditions is the same, greater, or less than it was a century ago. Or two centuries ago. Or a thousand years ago. In other words, to make the case for a declining humanity, we should see poverty not only keeping pace with population growth but those living in poverty even increasing as a percentage of the world's population.

Fortunately, when we take the time to examine the statistics carefully, we find some reason for optimism. According to an article in the Spring 2006 edition of the magazine *The International Economy*,[1] poverty rates declined approximately 4 percent every twenty years between 1820 (when economic statistics were first kept) and 1950. Between 1950 and 1980, that decline increased to a rate of 14 percent each twenty years, and even more impressive, since 1990 worldwide poverty rates have decreased an astonishing *20 percent*.

One example: in 1980, the poverty rates in India and China were 50 and 60 percent, respectively. By 2000, the poverty rates in both economies were in the range of 10 to 25 percent, meaning that the number of human beings who had moved out of poverty in these two countries alone was around a billion—a historical upliftment of *20*

---

1. Surjit S. Bhalla, "Today's Golden Age of Poverty Reduction." *The International Economy*, Spring 2006. This article can be found online at http://www.international-economy.com/Spring2006archive.htm (accessed June 23, 2009).

percent of the developing world's population! Clearly, the percentage of human beings living in poverty is going down (both proportionally and in actual numbers), while the world is seeing the birth of a small but growing middle class emerging in even the traditionally poorest countries. This is a fairly recent innovation in the human condition and a positive one.

Moreover, according to UN studies, world literacy rates stand at nearly 82 percent, with literacy rates approaching 99 percent in almost every industrialized country on the planet. Compare that with the worldwide literacy rate at the start of the twentieth century when, even after factoring in the industrialized nations (which were few and far between), literacy hovered at a dismal 5–10 percent. Today humanity is the most literate it has ever been at any point in its history—a trend that is likely only to continue to rise in the next century. If such trends continue, it's not unreasonable to imagine a worldwide literacy rate of 100 percent by the end of the twenty-first century.

Furthermore, in 1900 only about 10 percent of Americans had high-school diplomas. Today, only 10 percent *don't* have them, and there are twice as many institutions of higher education in the world today than there were a mere fifty years ago! In fact, in almost every area of human endeavor—infant mortality, life expectancy, annual income, access to medical attention, sanitation—the world has seen dramatic improvement in the last one hundred years. Of course, that's not to say that there aren't areas of the world still mired in abject poverty, or that AIDS and other diseases aren't cutting swaths through some populations, but compared to the quality of life experienced just a century ago by most human beings on the planet, ours is an infinitely better world today.

Human Rights and Civil Liberties

Consider also that human justice has improved considerably, especially over the last few centuries. Whereas slavery was once an acceptable practice around the world (and, in fact, played a major role in the development of civilization over the last seven thousand years), today it is illegal everywhere on the planet. Additionally, for the most part, people cannot be worked to death, imprisoned in a debtor's prison, or summarily executed for crimes as petty as stealing a loaf of bread or making off with another man's horse. Moreover, today people cannot be hanged for practicing witchcraft or burned at the stake for heresy, nor can a mob lynch a man because of the color of his skin and expect the judicial system to turn a blind eye as was usually the case a mere century ago.

The rights of women, children, and racial minorities have also improved substantially, especially when contrasted with the conditions in the past. Today, truly repressive societies, while still in existence, are growing less common and less sustainable. Of course, there are exceptions: child labor laws are sometimes lax in third-world countries and exploitation by the wealthy is still common (and, in places, prevalent), but the point is that such behavior, when exposed, is routinely prosecuted, whereas in the past such behavior was common, tolerated, and even expected. While we are still far from realizing a utopian world, it would be difficult to argue that humanity is not collectively becoming increasingly aware of the rights of all human beings, as well as becoming less willing to exploit the weak for the benefit of the wealthy.

Even our governmental institutions have demonstrated tremendous—and usually positive—changes in just the last sixty years. Immediately prior to the Second World War, there were just over two dozen functioning democracies on the planet; today, nearly two-thirds of the world's people live under some form of democracy[2] or benign, constitutional monarchy that guarantees individual liberties and human rights. Truly nihilistic, totalitarian governments can be counted on the fingers of one hand, while repressive societies—defined as those under the control of authoritarian leaders or those with dismal human rights records—number only in the dozens (out of a total of nearly two hun-

---

2. While admittedly some of these democracies are tentative at best and often corrupt, they do include some degree of representational government, making the people at least somewhat freer than they were during the authoritative regimes under which they had previously lived.

dred sovereign nations on the planet). Of course, there is always the danger of a major democracy failing and being replaced with an authoritative regime (as happened in Germany in 1933), but in a world in which each nation's economy is becoming increasingly integrated with those of its neighbors to form a genuinely global economy, such a descent into darkness would be more difficult to sustain.

Wars and Rumors of War

Largely as a result of this democratization process, war as an almost natural and expected instrument of foreign policy is growing increasingly unpopular and uncommon. Certainly the recent conflicts in Iraq and Afghanistan—regardless of how one feels about their necessity—are potent reminders to the world that war as a tool of foreign policy is indeed losing if not its effectiveness, then at least its luster. Though small-scale conflicts still rage in some isolated spots around the world (and are usually confined to nondemocratic nations), really big wars—that is, armed conflict between two or more sovereign nations—are increasingly rare, demonstrating that nations today are far less willing to resort to violence to resolve their disputes than they were in the past. Of course, the threat of terrorism or of a rogue nation developing and then actually using nuclear weapons remains, but these are mere flea bites when compared to the twin threats the world faced from Nazism and Communism throughout much of the twentieth century.

Consider also that, while the threat of nuclear war remains potent, the proliferation of nuclear weapons has dramatically decreased over the last two decades. In 1987 there were over 70,000 warheads of all types in the nuclear arsenals of the world's six nuclear powers. Today, due in a large part to various treaties negotiated in the 1980s and '90s and the collapse of the old Soviet empire (which held the lion's share of such warheads), the number is closer to 20,000, with further cuts anticipated in the coming years. While 20,000 warheads (about half of which are considered fully operational) is still a massive number, it's a far cry from the height of the Cold War, when the prospect of the destruction all life on the planet ten times over remained very real. Certainly, the fact that we are cutting our nuclear arsenals rather than increasing them has to be considered a positive sign for the future.

But what of the two great wars that ravaged the twentieth century, the horror of the Holocaust, Stalin's brutal gulags, and the more immediate concerns about rising crime rates, terrorism, and international drug trafficking? Couldn't it be said that modern history argues far more persuasively for a humanity in moral decline rather than for a world on the verge of a golden age of enlightenment?

The atrocities committed in the last hundred years *seem* so much worse than those in the past for two reasons. First is the fact that most people aren't aware of the many atrocities that took place in the past—when pogroms and the persecution of entire populations because of their religious or political views were common—making it appear that genocide is a modern aberration. Second, thanks to the media and the Internet, we are now more aware of the scope and horror of genocides when they are committed, whereas in the past such news seldom found its way into the popular consciousness. Additionally, when such atrocities are brought to light, they are now much more likely to incite an international outcry and response—something that was far less of a problem for a brutal regime only a few decades ago. But probably the biggest reason that mass murder appears to be more extensive today than in the past is because society has only relatively recently acquired the technology to systematically and efficiently kill very large numbers of people, which is what made possible events like the Holocaust or the killing fields of Cambodia. Without the technology of the gas chambers, the Holocaust would have been far less extensive, demonstrating that atrocities are not a testimony to humanity's growing violence but a demonstration of its growing technical acumen. Give Attila the Hun, Nero, Ivan the Terrible, or any of a host of other brutal leaders from antiquity the sort of technology available to us today, and they would have made the Holocaust look like child's play in comparison.

In the end, it's not whether there are more evil deeds and criminal acts being committed today than in the past, but whether there are *proportionately* more. No doubt more people die at the hands of their brothers today than they did in the past, but that's because there are more people than in the past. Naturally, when the population increases, the number of crimes is going to go up. That, however, is not necessarily indicative of an increase in frequency. For example, if a population were to double at the same time its crime rate fell by 40 percent, there would be still be a net increase in the total number of crimes committed, thereby giving the appearance that things were getting worse when statistically they are actually significantly improving. It's often just a numbers game.

Environmental Threats

But certainly the destruction of our environment argues strongly for a planet in peril, does it not? With the world's population approaching seven billion; the threat posed to our weather, oceans, and ecosystems by climate change; the tragic consequences of deforestation; and the reality of rapidly dwindling energy supplies and reduced sources of new energy to feed an ever more ravenous and growing world economy, how can anyone even begin entertaining the possibility that things are improving in any way?

No one said there aren't challenges, but consider that the first step in overcoming adversity is becoming aware that a threat exists in the first place. For decades, the world went about its business creating an industrial revolution the likes of which the planet had never experienced before, without giving a second thought to the consequences of spewing clouds of noxious gas into the environment, clear-cutting entire forests, and filling lakes and rivers with all kinds of toxic pollutants. Now, however, we are taking steps to change all that. Emissions standards for both factories and automobiles are now strictly enforced; logging is heavily regulated; species once on the verge of extinction have been brought back to viability; recycling is one of the fastest growing industries on the planet; and the push toward clean, renewable energy is well underway.

We obviously have a long way yet to go, and unrestricted growth among rapidly industrializing countries like China and India remains a concern (as it does within already industrialized nations in the West), but for the most part an effort is underway to rein in the worst of the abuses. Just the mere fact that we are aware of the magnitude of the problem, and are willing to pursue solutions, can't be seen as anything but positive. Undoubtedly, the world has a number of challenges ahead and many political, economic, and technological hurdles to overcome—along with a few unforeseen catastrophes to endure—but a very good case can be made that we are on the right track.

Conclusions

Those who hold to the disintegrating-society theory are unlikely to be impressed with these examples, but I submit that the gloomy outlook of such people can only be sustained if they determinedly ignore the judgment of history. While in many ways we are still a very brutal people who are but one mindless act away from exterminating ourselves, we are at the same time far less tolerant than we used to be of those who diminish the value of human life, rape the environment, or practice injustice. Perhaps this is the result of numerous reincarnations, or maybe it's simply that we're naturally—albeit slowly—advancing as a species (or perhaps it's a case of the one leading to the other). Whatever the cause, the case for increasing spirituality and enlightenment on a political, social, and religious level around the world can be denied only if one insists on keeping both eyes firmly shut.

It's not a perfect world that we seek, nor is such a world even possible. However, we can create a planet on which humanity respects itself and its environment, where we solve problems through dialogue and hard work, and where each day brings us closer to realizing the potential that resides within us. It may not be a utopia in the classic sense, but it would be enough to lay the foundation for an exciting future among the stars that is, in my opinion, not only our destiny but our birthright. It may not happen in my lifetime, or within the lifetime of my readers, but it is a certainty we will one day obtain, no matter how long it takes. We are the authors of our destiny—not fate, not the stars, and not even the gods. I believe we are just beginning to understand that, and once we fully appreciate it we will find the strength within ourselves to create the sort of world of which we can be proud.

Naïve? Perhaps, but it's a naïveté I am only too happy to embrace.

Conclusion

Today I look back at my dalliance with end-times beliefs and wonder why I was so quick to believe Hal Lindsey and his spectacular doomsday scenarios, and even more curiously, why I held on to those beliefs as long as I did. I suppose it was all part of a learning process.

But more than that, my experiences gave me an insight into the world of doomsday that I could never begin to appreciate had I not walked down that path. And certainly this book could never have been written had I remained above the fray, confident in my own superior discernment skills as so many skeptics are.

Most of all, I think reading Lindsey's book and the subsequent experiences I went through helped me understand a flaw in humanity—and myself—that needs to be addressed before we can move on as a species, and that is the role fear plays in all our lives. I never realized when I was a boy how much that basic, raw emotion controls us; dictates what we do, think, believe, and feel; and generally shapes our perceptions of the world around us. It is fear that turns our planet into a sometimes scary place, and doomsday prophecies are the natural byproduct of that fear.

Perhaps therein lies the real tragedy of doomsday beliefs: they rob us of our optimism about the future, leaving us with a world largely bereft of hope, where fear is the norm and despair the only reasonable response to it all. They also destroy our incentive to work toward overcoming the obstacles fate places before us and to use our imagination to come up with practical solutions to whatever crises we may face—obstacles, I believe, that are placed there *precisely so we might grow as a species*. And, finally, such beliefs rob humanity of its meaning, for if we are all to be pulverized into dust in the next few years, decades, or centuries, what does this experiment known as *Homo sapiens* mean? Wasn't it all one great waste of time and potential?

I don't believe it is. While I am perfectly willing to acknowledge that we live in a world fraught with danger, I believe we also live in one filled with the promise of a potential golden age if we can only overcome the fears that prevent us from moving toward that destiny. It won't be easy, but I suspect that someday we will look back at this primitive era as one in which humanity first "saw the light" and began moving past its fear, and in that may lie the true test of humanity and the best evidence yet of its inherent divinity.

Fortunately, there is hope. We have reason to reconsider our dire future, especially in the perfect hindsight of history. Our forebears may not have had global pandemics and melting icecaps to worry about, but they were just as certain of the end as we appear to be today. They, however, were wrong in their suppositions, just as I believe we will be about ours.

Regardless of the reasons we are so quick to grasp on to doomsday scenarios, it is my contention that they are unhealthy on both a societal and personal level. They are essentially designed to frighten us, to have us live our lives certain that every world event, every weather anomaly, and every terrorist attack is but the first hoofbeats of the apocalypse, which is a difficult way to live indeed. That's not to suggest that there aren't genuine dangers out there we need to be concerned with; the potential for nuclear terrorism, environmental catastrophe, rampant overpopulation, and social unrest are very real problems we do need to confront. However, we can't appreciate the degree of danger they truly represent as long as we live in a world awash in sensationalistic and pseudoscientific doomsday scenarios that serve only to distract us from the real problems we face.

While admittedly most such musings are essentially harmless, there is a dark side to doomsday scenarios that needs to be appreciated, which is the fact that what you believe *does* impact your life, often in important ways. As the thirty-nine men and women of the Heaven's Gate cult and the 913 men, women, and children of the Jonestown suicide pact demonstrate, the consequences for believing in doomsday can be quite serious.

But if we can learn to understand why we accept these scenarios in the first place and how they work in shaping our perceptions about the world around us, we can come to appreciate the potential damage such beliefs can pose and therefore protect ourselves from the real dangers many of the more extreme end-times teachings contain within them. So how can we protect ourselves from being victimized by such ideas? It's not difficult, really: simply maintain a degree of skepticism when such claims are made, insist on supporting data whenever possible, examine counterarguments when presented, and recognize your own capacity to be deceived. There's nothing wrong with admitting that you might be a victim of your own enthusiasm; in fact, the ability to admit such is a sign of intellectual flexibility rather than a flaw in character.

We have all believed things in the past that we know now to have been erroneous, and we put them behind us when we discover them—if we are intellectually honest with ourselves and emotionally mature enough to do so. Additionally, all of us believe things today that we will one day discover to be untrue, probably to our initial embarrassment but eventually to our greater understanding and improved discernment. We all fall for fakes and nonsense sometimes; it is a part of what it means to be human. Recognize your own fallibility, however, and the doomsday prophets cannot entice you into their world of fear and darkness, no matter how compelling their predictions or how sincere they may seem.

Such is the start of wisdom.

Free Catalog

Get the latest information on our body, mind, and spirit products! To receive a **free** copy of Llewellyn's consumer catalog, *New Worlds of Mind & Spirit,* simply call 1-877-NEW-WRLD or visit our website at www.llewellyn.com and click on *New Worlds.*

LLEWELLYN ORDERING INFORMATION

Order Online:
Visit our website at www.llewellyn.com, select your books, and order them on our secure server.

Order by Phone:
- Call toll-free within the U.S. at 1-877-NEW-WRLD (1-877-639-9753). Call toll-free within Canada at 1-866-NEW-WRLD (1-866-639-9753).
- We accept VISA, MasterCard, and American Express

Order by Mail:
Send the full price of your order (MN residents add 6.875% sales tax) in U.S. funds, plus postage & handling to:

> Llewellyn Worldwide
> 2143 Wooddale Drive, Dept. 978-0-7387-1464-6
> Woodbury, MN 55125-2989

Postage & Handling:

Standard (U.S., Mexico & Canada). If your order is:
  $24.99 and under, add $4.00
  $25.00 and over, FREE STANDARD SHIPPING

AK, HI, PR: $16.00 for one book plus $2.00 for each additional book.

International Orders (airmail only):
  $16.00 for one book plus $3.00 for each additional book

Orders are processed within 2 business days.
Please allow for normal shipping time. Postage and handling rates subject to change.

Atlantis: Lessons from the Lost Continent
J. Allan Danelek

Is the Atlantis story a myth, pseudo-science, or a true story with lessons for our future? Objective and scrupulous, J. Allan Danelek applies his signature no-nonsense approach to the legend of the Lost Continent.

Investigating Plato's dialogues, geosciences, traditional theories, and historical maps, Danelek attempts to answer the questions surrounding this twelve-thousand-year-old legend. Did Atlantis truly exist? If so, what was its culture like? How did the Atlanteans destroy themselves? Why haven't we found any evidence of this civilization? And finally, what can we learn from the fate of Atlantis—an advanced civilization perhaps not unlike our own?

This engaging exploration of Atlantis brings levity to a controversial subject and offers reasonable and fascinating theories of what may have happened to this ancient civilization.

978-0-7387-1162-1, 264 pages $15.95

To order, call 1-877-NEW-WRLD
Prices subject to change without notice
Order at Llewellyn.com 24 hours a day, 7 days a week!

Beyond 2012

A Shaman's Call to Personal Change and the Transformation of Global Consciousness

James Endredy

War, catastrophic geologic events, Armageddon . . . the prophecies surrounding 2012—the end of the Mayan calendar—aren't pretty. James Endredy pierces the doom and gloom with hope and a positive, hopeful message for humankind.

For wisdom and guidance concerning this significant date, Endredy consults Tataiwari (Grandfather Fire) and Nakawe (Grandmother Growth)—the "First Shamans." Recorded here is their fascinating dialog. They reveal how the evolution of human consciousness, sustaining the earth, and our personal happiness are all interconnected.

Discover what you can do to spur the transformation of human consciousness. See how connecting with our true selves, daily acts of compassion and love, focusing personal energy, and even gardening can make a difference. Endredy also shares shamanistic techniques to revive the health of our planet . . . and ourselves.

978-0-7387-1158-4, 240 pages $16.95

To Write to the Author

If you wish to contact the author or would like more information about this book, please write to the author in care of Llewellyn Worldwide and we will forward your request. Both the author and publisher appreciate hearing from you and learning of your enjoyment of this book and how it has helped you. Llewellyn Worldwide cannot guarantee that every letter written to the author can be answered, but all will be forwarded. Please write to:

J. Allan Danelek
℅ Llewellyn Worldwide
2143 Wooddale Drive, Dept. 978-0-7387-1464-6
Woodbury, MN 55125-2989, U.S.A.
Please enclose a self-addressed stamped envelope for reply,
or $1.00 to cover costs. If outside the U.S.A., enclose
an international postal reply coupon.

Many of Llewellyn's authors have websites with additional information and resources. For more information, please visit our website at http://www.llewellyn.com.